W9-AGV-455

Multivariate Approximation

Multivariate Approximation

Edited by

D. C. HANDSCOMB

Oxford University Computing Laboratory
Oxford, England

1978

ACADEMIC PRESS
London · New York · San Francisco
A Subsidiary of Harcourt Brace Jovanovich, Publishers

ACADEMIC PRESS INC. (LONDON) LTD.
24/28 OVAL ROAD,
LONDON NW1

U.S. Edition published by
ACADEMIC PRESS INC.
111 FIFTH AVENUE
NEW YORK, NEW YORK 10003

Library of Congress Catalog Card Number: 78–18022

ISBN: 0–12–323350–X

Printed in Great Britain by
Whitstable Litho Ltd., Whitstable, Kent.

LIST OF CONTRIBUTORS

M.F. BARNSLEY, Service de Physique Théorique, C.E.N.S., B.P. No.2 - 91190 Gif-sur Yvette, France.

H.-P. BLATT, Fakultät für Mathematik und Informatik, Universität Mannheim, 68 Mannheim A5, Germany.

K.R. BUTTERFIELD, Atkins Research and Development, Ashley Road, Epsom, Surrey, England.

E.W. CHENEY, Department of Mathematics, University of Texas at Austin, Austin, Texas, U.S.A.

J.S.R. CHISHOLM, Mathematical Institute, Cornwallis Building, University of Kent, Canterbury CT2 7NF, England.

L. COLLATZ, Institut für Angewandte Mathematik, Universität Hamburg, 2 Hamburg 13, Bundesstrasse 55, Germany.

F.J. DELVOS, Fachbereich 6 Mathematik I, Gesamthochschule Siegen, Hölderlinstr. 3, D-5900 Siegen 21, Germany.

S.W. ELLACOTT, Department of Mathematics, Brighton Polytechnic, Moulsecoomb, Brighton BN2 4GJ, England.

A.R. FORREST, School of Computing Studies, University of East Anglia, Norwich NR4 7TJ, England.

J.A. GREGORY, Department of Mathematics, Brunel University, Uxbridge UB8 3PH, England.

J.G. HAYES, Division of Numerical Analysis and Computing, National Physical Laboratory, Teddington TW11 0LW, England.

L.A. KARLOVITZ, National Science Foundation, Washington D.C. 20550, U.S.A.

G.G. LORENTZ, University of Texas at Austin, Department of Mathematics RLM 8-100, Austin, Texas 78712, U.S.A.

J.C. MASON, Mathematics Branch, Royal Military College of Science, Shrivenham, Swindon SN6 8LA, England.

G. MEINARDUS, Fachbereich 6 Mathematik IV, Gesamthochschule Siegen, Hölderlinstr. 3, D-5900 Siegen 21, Germany.

J. MEINGUET, Institut de Mathématique P. et A., Université de Louvain, Chemin du Cyclotron 2, B-1348 Louvain-la-Neuve, Belgium.

E.L. ORTIZ, Department of Mathematics, Imperial College, Huxley Building, Queen's Gate, London SW7 2BZ, England.

GEETHA S. RAO, Ramanujan Institute, University of Madras, Madras, India.

J.R. RICE, Computer Sciences, Mathematical Sciences Building, Purdue University, West Lafayette, Indiana 47907, U.S.A.

T.J. RIVLIN, Mathematical Sciences Department, IBM T.J. Watson Research Center, Yorktown Heights, N.Y. 10598, U.S.A.

H.S. SHAPIRO, Institutionen för Matematik, Kungl. Tekniska Högskolan, Stockholm 70, Sweden.

D.D. STANCU, Faculty of Mathematics, University of Cluj, Cluj, Romania.

H. STRAUSS, Institut für Angewandte Mathematik, Universität Erlangen-Nürnberg, 852 Erlangen, Martenstr. 3, Germany.

H. WALLIN, Institute of Mathematics and Statistics, University of Umeå, S-901 87 Umeå, Sweden.

G.A. WATSON, Department of Mathematics, The University, Dundee DD1 4HN, Scotland.

PREFACE

The theory of approximation has a long history, of which a notable feature is the surge of interest in the 1960's, leading to the publication at that time of a number of textbooks and the launching in 1968 of a specialist journal, the Journal of Approximation Theory. The subject is now regularly included in courses of numerical, classical or functional analysis, while the progress of research has been such that we can tell the man in the drawing office almost anything he wants to know about fitting curves to points in a plane or representing functions of one variable.

The situation with regard to surface fitting, or the representation of functions of several variables, is quite different; here the subject is still in its infancy. There are, it is true, results in abstract approximation theory that could be considered as falling as much into this field as into any other, but if one looks at the contents of a typical volume of the above-mentioned journal, or of the proceedings of one of the frequent conferences on approximation held in recent years, it may surprise one to realise how little specific mention there is of functions of more than one variable. Yet the potential applications of multivariate approximation to real-world problems are more varied, and could well be more rewarding, than those of univariate approximation.

Consequently when in 1975 the London Mathematical Society conceived the idea, as part of a continuing programme of research symposia on various branches of mathematics, of including a meeting devoted to approximation theory, it was felt that, rather than duplicate the kind of activity that was already well catered for, it would be useful to pay some specific attention to this neglected facet. Accordingly a committee was set up, and invited a number of persons believed to have an active interest in some form of multivariate approximation to take part in a symposium, to be held at the University of Durham from July 21 to July 30, 1977.

In the event, including others who wrote in asking to attend, the Symposium was an international gathering of 48 participants, some of them accompanied by wives and families. (Here we must mention our disappointment that Ivan Singer, who had been invited to come as one of the principal speakers, was at the last minute prevented by lack of a visa from leaving Romania.) A programme of invited and contributed lectures was arranged, most of which appear in this volume. It must be realised, however, that these lectures account for only a part of the available time, and that at least as much of the value of the symposium lay, as was always the intention, in the lively discussions following the lectures and the informal contacts made at other times, none of which are recorded here. Neither should the reader look here for a unity of approach or an all-inclusive coverage; all that we have consciously attempted is to bring together speakers from diverse theoretical and practical backgrounds. We were unable to raise a speaker from the Soviet Union, so that an account of recent work in that country is possibly the most outstanding omission from this book.

The success that was generally felt to attend the Symposium must be ascribed in the first place to those who spoke and those who came to listen. The Symposium was housed with efficient yet relaxed hospitality in Grey College, Durham, and was supported financially by generous grants from the London Mathematical Society and from the European Research Office of the United States Army, to all of whom it is a pleasure to record our debt of gratitude. Finally I must personally thank my fellow organisers, Professors Roy Chisholm, Heini Halberstam and Alan Talbot, as well as Professor Tom Willmore and Dr Robert Johnson of the University of Durham.

June 1978 DAVID HANDSCOMB

CONTENTS

SOME ASPECTS OF OPTIMAL RECOVERY

T.J. Rivlin

Mathematical Sciences Department
IBM T.J. Watson Research Center
Yorktown Heights, N.Y. 10598

1. INTRODUCTION

By optimal recovery we mean optimal estimation of a function known to belong to a given set of functions using limited further information about it. In [1] Micchelli and Rivlin gave a general framework and some theory for such problems, as well as many examples. The explicit examples presented there involved functions of one variable. Our purpose here is to re-examine some of that material as a guide to some preliminary observations about optimal recovery in the realm of multivariate functions.

Let us look at a typical example. Suppose a function $f \in C^{(n-1)}[0,1]$ satisfies $|f^{(n-1)}(t_1) - f^{(n-1)}(t_2)| \le |t_1-t_2|$ for all $t_1, t_2 \in [0,1]$ and some $n \ge 1$. Call the set of such functions A_n. Suppose further that in addition to knowing that $f \in A_n$ we also know $f(x_1),\ldots,f(x_{n+r})$ where $x_1,\ldots,x_{n+r}$ are given and satisfy $0 \le x_1 < x_2 < \ldots < x_{n+r} \le 1$. The problem is to obtain the best possible estimate of, say, $f(\tau)$ using the sampled values $f(x_1),\ldots,f(x_{n+r})$ and the fact that $f \in A_n$. To be precise: let α be any function whose domain is the subset of $\mathbb{R}^{n+r}$ consisting of vectors $(f(x_1),\ldots,f(x_{n+r}))$ for any $f \in A_n$, and whose range is in $\mathbb{R}$. Then

$$E(\alpha) = \sup_{f\in A_n} |f(\tau) - \alpha(f(x_1),\ldots,f(x_{n+r}))|$$

is the *error of the algorithm* α , and

$$E^* = \inf_{\alpha} E(\alpha)$$

is called the *intrinsic error* in the problem. If there exists an algorithm α^* such that $E(\alpha^*) = E^*$ then α^* is an *optimal algorithm* and provides an optimal recovery of $f(\tau)$.

For example, consider the case of $r = 0$. The polynomial $Q(t) = (t-x_1)\ldots(t-x_n)/n!$ is an element of A_n , as is $-Q(t)$. If α is any algorithm then

$$|Q(\tau) - \alpha(Q(x_1),\ldots,Q(x_n))| \le E(\alpha)$$

and

$$|-Q(\tau) - \alpha(-Q(x_1),\ldots,-Q(x_n)| \le E(\alpha) \quad .$$

Thus since $Q(x_i) = 0$, $i = 1,\ldots,n$ we see that $E(\alpha) \ge |Q(\tau)|$ and hence $E^* \ge |Q(\tau)|$. Moreover, if $\alpha^*(f(x_1),\ldots,f(x_n)) = P(\tau)$, where P is the unique polynomial of degree at most $n-1$ which agrees with f at $x_1,\ldots,x_n$, then for every $f \in A_n$

$$|f(\tau) - \alpha^*(f(x_1),\ldots,f(x_n))| = |f(\tau) - P(\tau)| \le |Q(\tau)| \quad .$$

For, if $f(\tau) - P(\tau) = \lambda Q(\tau)$ where $|\lambda| > 1$, then Rolle's Theorem implies that $f^{(n-1)}(t) - P^{(n-1)}(t) - \lambda Q^{(n-1)}(t) = h(t)$ has two distinct zeros in $[0,1]$. Since $f \in A_n$ and $P^{(n-1)}(t) = \lambda Q^{(n-1)}(t)$ is a linear polynomial whose slope is λ, we see that $h(t)$ is strictly monotone in $[0,1]$ and so $h(t)$ has at most one zero in that interval. Thus $E(\alpha^*) \le |Q(\tau)| \le E^*$, and by the definition of E^*, $E^* = |Q(\tau)|$ and polynomial interpolation is an optimal algorithm.

For any $r \ge 0$ it is shown in Micchelli, Rivlin and Winograd[5] that there exists a unique (up to multiplication by -1) perfect spline, $q(t)$, of degree n having r knots $\zeta_1,\ldots,\zeta_r$ which satisfy $0 < \zeta_1 < \ldots < \zeta_r < 1$ such that:

i) $q(x_i) = 0$, $i = 1,\ldots,n+r$

and

ii) $|q^{(n)}(t)| \equiv 1$, $t \ne \zeta_i$, $i = 1,\ldots,r$.

This perfect spline is the generalisation of its namesake in the case $r = 0$. $|q(\tau)|$ is the intrinsic error, and an optimal algorithm is obtained by interpolation to $f(x_1),\ldots, f(x_{n+r})$ by a spline of degree $n-1$ having as knots precisely $\zeta_1,\ldots,\zeta_r$. That is, if $s_f(t)$ is the unique spline of degree $n-1$ having knots $\zeta_1,\ldots,\zeta_r$ and satisfying $s_f(x_i) = f(x_i)$, $i = 1,\ldots,n+r$ then $|f(\tau) - s_f(\tau)| \le |q(\tau)|$ for every $f \in A_n$ and $\alpha^*(f(x_1),\ldots,f(x_{n+r})) = s_f(\tau)$ defines an optimal algorithm. Note that the knots of q are independent of τ. Thus if our object were to recover the function f (rather than its value at a preassigned point) in the sense of producing as good a uniform approximation on $[0,1]$ as possible from the samples $f(x_1),\ldots,f(x_{n+r})$, the intrinsic error must be $\max|q(\tau)|$, $\tau \in [0,1]$ and the above-mentioned spline interpolation provides an optimal algorithm if we ask only that the approximation take place in $L^\infty[0,1]$. For details see also Micchelli and Rivlin[4], Gaffney and Powell[3] and Bojanov and Chernogorov[1].

In general the universal knots $\zeta_1,\ldots,\zeta_r$ can only be obtained approximately (see de Boor[2]), but in the case $n = 1$ the sketch of $q(t)$ in Fig. 1 shows that $\zeta_i = (x_i+x_{i+1})/2$, $i = 1,\dots,r$.

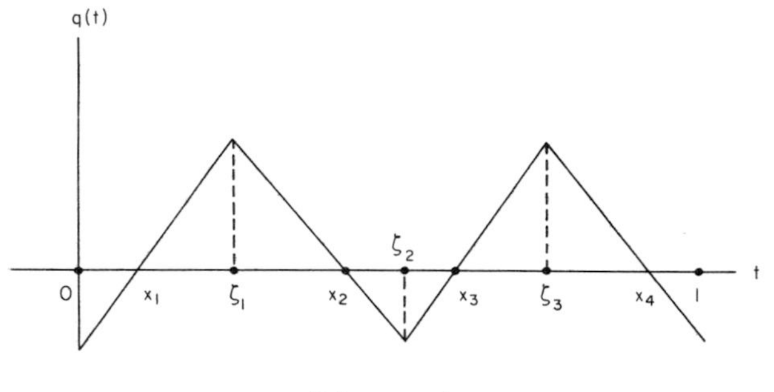

Figure 1

Note that an optimal algorithm is now obtained by mapping $(f(x_1),\dots,f(x_{n+r}))$ to the value at τ of the step-function having the value $f(x_i)$ in $[\zeta_{i-1},\zeta_i)$, $i = 1,\dots,r$ and $f(x_{r+1})$ in $[\zeta_r,1]$, where $\zeta_o = 0$.

2. A GENERAL THEORY

Micchelli and Rivlin[4] give the following general framework for the optimal recovery problem. The object, x, about which we have some limited information, is taken to be an element of a subset K of a linear space X. Our aim is to recover Ux, where U is a linear operator into a normed linear space Z. A linear operator, I, the information operator, maps X into the normed linear space Y. However, we assume that Ix for $x \in K$ may be contaminated by error. That is, we actually know only some $y \in Y$ satisfying $\|Ix-y\| \le \varepsilon$, where $\varepsilon \ge 0$ is some preassigned tolerance. (In the introductory example, and in the further examples we are assuming that $\varepsilon = 0$). An algorithm is now taken to be any function α whose domain is $IK + \varepsilon S$ (where S is the unit ball in Y) and whose range is in Z . A diagram of the process is given in Fig. 2.

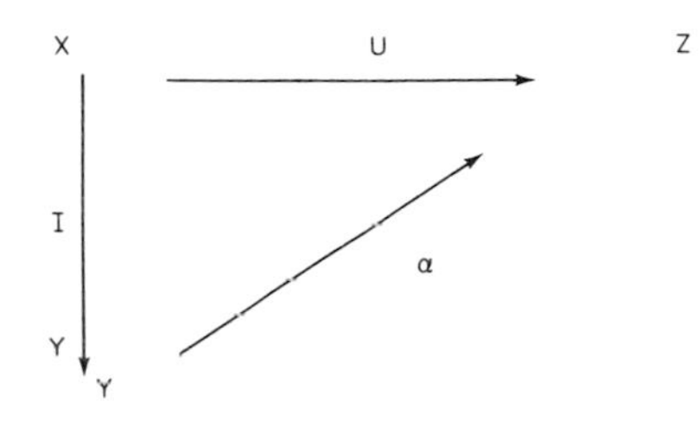

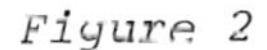

Figure 2

$$E(\alpha) = \sup_{\substack{x \in K \\ \|Ix-y\| \leq \varepsilon}} \|Ux - \alpha y\|$$

is the error of α and

$$E^* = \inf_{\alpha} E(\alpha)$$

is the intrinsic error. If $E(\alpha^*) = E^*$, α^* is called an optimal algorithm.

It is easy to show, by the same argument we used in the polynomial case of the introductory example, that if K is a convex and balanced subset of X ,

$$E^* \geq \sup_{\substack{x \in K \\ \|Ix\| \leq \varepsilon}} \|Ux\| \quad . \tag{2.1}$$

Indeed, equality actually holds in (2.1) in many interesting cases. Many more details are given in Micchelli and Rivlin[4].

3. ANALYTIC FUNCTIONS

In order to test our wings for multivariable flight let us look at some examples of the general structure we have just erected involving analytic functions in the plane.

Take X to be H^∞, the space of bounded analytic functions in the unit disc $D: |z| < 1$. If $f \in X$ let $\|f\| = \sup |f(z)|$, $z \in D$, and choose $K = \{f \in H^\infty : \|f\| \le 1\}$. Suppose $z_1, \ldots, z_n$ are given points of D, $If = (f(z_1), \ldots, f(z_n))$ (so that $Y = C^n$) and $\varepsilon = 0$. Take $Z = C$ and suppose $Uf = f(\zeta)$ where ζ is some given point of D.

This is a fairly simple optimal recovery problem. Put

$$B_n(z) = \prod_{i=1}^{n} \frac{z - z_i}{1 - \bar{z}_i z}$$

If $f \in K$ and $If = 0$ then $(f/B_n) \in K$ and by the maximum principle $|f(\zeta)| \le |B_n(\zeta)|$. Since $B_n \in K$ and $IB_n = 0$ we have

$$\sup_{\substack{\|f\| \le 1 \\ f(z_i) = 0}} |f(\zeta)| = |B_n(\zeta)| \le E^* .$$

Indeed, the general theory alluded to in the preceding section tells us that $E^* = |B_n(\zeta)|$, but we shall establish that fact independently in a moment.

But if $\alpha^*(f(z_1), \ldots, f(z_n)) = \alpha_1^* f(z_1) + \ldots + \alpha_n^* f(z_n)$ is determined, using the calculus of residues, by

$$f(\zeta) - \sum_{i=1}^{n} \alpha_i^* f(z_i) = \frac{1}{2\pi i} = \int_{|z|=1} \frac{B_n(\zeta)}{B_n(z)} \frac{1 - |\zeta|^2}{1 - z\bar{\zeta}} \frac{1}{z - \zeta} f(z)\, dz$$

then

$$E(\alpha^*) \le (|B_n(\zeta)|) \frac{1}{2\pi} \int_0^{2\pi} \frac{1-|\zeta|^2}{|e^{i\theta}-\zeta|^2} d\theta = |B_n(\zeta)| \quad ,$$

$E^* = E(\alpha^*) = |B_n(\zeta)|$ and α^* is an optimal algorithm.

The optimal recovery of $f'(\zeta)$ under the same assumptions and the optimal recovery of $f(\zeta)$ when $K = \{f \in H^\infty : \|f'\| \le 1\}$ and $z_1 = \ldots = z_n = 0$ are described in Micchelli and Rivlin[4].

4. FUNCTIONS OF TWO VARIABLES

(4.1) Let B denote the disc $|z| \le 1$ and suppose $\zeta: (\xi,\eta) \in B$, as are the points $z_i: (x_i,y_i)$, $i = 1,\ldots,n$. Let $X = L^\infty(B)$ be the real-valued essentially bounded functions on B, and K be the subset of $f \in X$ satisfying $|f(z)-f(z')| \le |z-z'|$. Our objective is to recover $f(\zeta)$ from the exact data $f(z_1),\ldots,f(z_n)$. Put $K_o = \{f \in K: f(z_i) = 0,\ i = 1,\ldots,n\}$ and let k be any index such that

$$\rho = \min_j |\zeta - z_j| = |\zeta - z_k| \quad .$$

Consider

$$g_\zeta(z) = g(z) = \begin{cases} \rho - |\zeta - z| & , \quad |\zeta - z| \le \rho \\ 0 & , \quad |\zeta - z| \ge \rho \end{cases}$$

for $z \in B$. Clearly $g(\zeta + re^{i\phi}) = (\rho - r)_+$ and if $\zeta_1 = \zeta + r_1 e^{i\phi}$ and $\zeta_2 = \zeta + r_2 e^{i\phi_2}$ are any two points of B, $g(\zeta_1) - g(\zeta_2) = (\rho - r_1)_+ - (\rho - r_2)_+$ and hence

$$|g(\zeta_1) - g(\zeta_2)| \le |r_1 - r_2| \le |\zeta_1 - \zeta_2| \quad .$$

Thus $g \in K_o$ and

$$E^*_\zeta \ge \sup_{f \in K_o} |f(\zeta)| \ge |g(\zeta)| = \rho \quad .$$

But the algorithm defined by $\alpha^*(f(z_1),\ldots,f(z_n)) = f(z_k)$ satisfies

$$\sup_{f\in K} |f(\zeta) - \alpha^* If| = \sup_{f\in K} |f(\zeta) - f(z_k)| \le |\zeta - z_k| = \rho \quad ,$$

hence $E^*_\zeta = \rho$ and α^* is optimal.

Suppose now that we want to recover f globally (that is, take $Z = X$ and U to be the identity operator) from $f(z_1),\ldots,f(z_n)$. The intrinsic error now satisfies

$$E^* \ge \max_{\zeta\in B} \min_j |\zeta - z_j| \quad . \tag{4.1.1}$$

To establish this let

$$B_k = \{\zeta\in B\colon \min_j |\zeta - z_j| = |\zeta - z_k|\} \ , \ k - 1,\ldots,n \quad .$$

For each k , B_k is a closed convex subset of B and every $\zeta \in B$ is in some B_k . If, now, following our previous notation, we put $q(z) = |z-z_k|$, $z \in B_k$, $k = 1,\ldots,n$, then we claim that $q \in K_o$. For suppose $u,v \in B$, $u \in B_i$, $v \in B_j$ and $q(v) \ge q(u)$, say. Then $q(v) = |v-z_j| \le |v-z_i|$ and hence $|q(v) - q(u)| = q(v) - q(u) \le |v-z_i| - |u-z_i| \le |v-u|$. Hence

$$E^* \ge \|q\| = \max_{z\in B} |q(z)| = \max_{\zeta\in B} \min_j |\zeta - z_j| \quad .$$

If we set $s(z) = f(z_k)$ on int B_k then $\alpha^*(f(z_1),\ldots, f(z_n)) = s(z)$ defines an optimal algorithm and $E^* = \|q\|$. For on each int B_k we have $|f(z) - s(z)| = |f(z) - f(z_k)| \le |z-z_k| \le \|q\|$.

Remark.

If we wish to recover $Uf = \int_B f$ from $f(z_1),\dots,f(z_n)$ for $f \in K$, then since $q \in K_o$ the intrinsic error satisfies

$$E^* \geq \left|\int_B q\right| = \int_B q \quad . \tag{4.1.2}$$

But for every $f \in K$ and every $z \in B$ we just saw that

$$-q(z) \leq f(z) - s(z) \leq q(z) \quad ,$$

and hence

$$\left|\int_B f - \int_B s\right| \leq \int_B q \leq E^* \quad .$$

Thus equality holds in (4.1.2) and

$$\sum_{i=1}^{n} b_i f(z_i)$$

is an optimal quadrature formula, where b_i is the area of B_i , $i = 1,\dots,n$.

(4.2) A problem of considerably greater difficulty and interest is the following. Let X and K be as in the beginning of Section 4.1, but now suppose that our aim is to recover $f(\zeta)$ from information presented in the form of line integrals along a finite number of chords of B . For example: Take $If = (I_{ij})$ where

$$I_{ij} = \int_{L_{ij}} f\,ds\,, \quad i = 0,\pm 1,\dots,\pm m \; ; \; j = 0,1,\dots,n-1 \quad ,$$

L_{ij} being the chord of B whose "distance" from the origin is ih (h is a given positive quantity) and whose normal makes an angle $(j\pi)/n$ with the positive x-axis. (See Fig.3.)

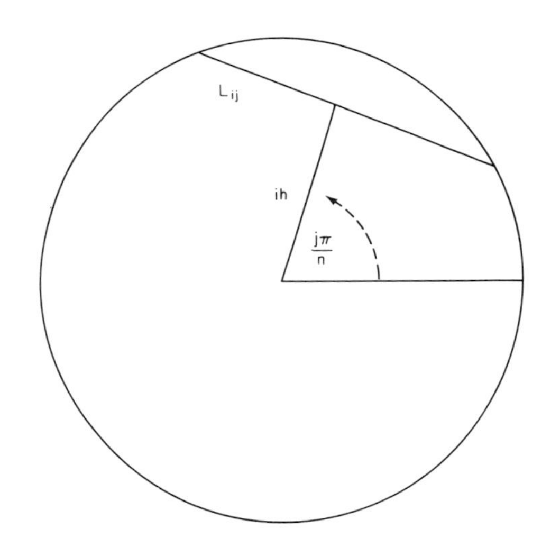

Figure 3

The problem of recovering a function from information of this kind is a model of reconstruction from X-rays of a cross-section of the head. However, in a realistic model of the X-ray problem the specification of X, K and U seems difficult (cf. Smith, Wagner and Guenther[6]). We conclude by treating the most elementary example of such a problem:

Given

$$\lambda = \frac{1}{2}\int_{-1}^{1} f(x,0)\,dx$$

what is the best estimate of $f(\zeta)$ where $\zeta \in B$ if $f \in K$?

Note that

$$|f(\xi,\eta) - \lambda| = \frac{1}{2}\left|\int_{-1}^{1} [f(\xi,\eta) - f(x,0)]dx\right|$$

$$\leq \frac{1}{2}\int_{-1}^{1} [(\xi-x)^2 + \eta^2]^{1/2}dx = c \ .$$

But

$$g(z) = c - |\zeta - z|$$

is in K_o and $g(\zeta) = c$. Therefore the estimate λ for $f(\zeta)$ is optimal and c is the intrinsic error.

REFERENCES

1. Bojanov, B.D. and Chernogorov, V.G. (1977). An optimal interpolation formula, *J. Approximation Theory* 20, 264-274.

2. de Boor, C. (1977). Computational aspects of optimal recovery, *in* "Optimal Estimation in Approximation Theory" (Eds. C.A. Micchelli and T.J. Rivlin), 69-91, Plenum Press, New York.

3. Gaffney, P.W. and Powell, M.J.D. (1976). Optimal interpolation, *in* "Numerical Analysis" (Ed. G.A. Watson) Lecture Notes in Mathematics, Vol.506, Springer, Heidelberg.

4. Micchelli, C.A. and Rivlin, T.J. (1977). A survey of optimal recovery, *in* "Optimal Estimation in Approximation Theory" (Eds. C.A. Micchelli and T.J. Rivlin), 1-54, Plenum Press, New York.

5. Micchelli, C.A., Rivlin, T.J. and Winograd, S. (1976). The optimal recovery of smooth functions, *Numer. Math.* 26, 191-200.

6. Smith, K.T., Wagner, S.L. and Guenther, R.B. (1976). Reconstruction from X-rays, *in* "Optimal Estimation in Approximation Theory" (Eds. C.A. Micchelli and T.J. Rivlin), 215-227, Plenum Press, New York.

APPLICATION OF MULTIVARIATE APPROXIMATION TO THE SOLUTION OF BOUNDARY VALUE PROBLEMS

L. Collatz

Institut für Angewandte Mathematik
Universität Hamburg
Hamburg, Germany

SUMMARY

Multivariate approximation occurs in many fields of applications, for instance in problems with linear and nonlinear partial differential equations, in integral equations, a.o. Examples are given for these types of problems. An important aid for the criteria, how good the approximation is, are the H-sets. H-sets are constructed for linear and nonlinear cases. Numerical examples illustrate the applicability of the theory.

1. INTRODUCTION

Multivariate approximation has similar applications as the onedimensional approximation, for instance to substitute a given complicated function of several variables by simpler functions the computer can easily calculate. Other applications occur in solving differential - and integral - and more general functional equations and here a very important case is the solution of partial differential equations. Therefore multivariate approximation is very important, because approximation methods are in many cases the only ones which give exact error bounds for the approximate solutions of the boundary value problems. Unfortunately one knows very little about the often unusual types of multivariate approximations which occur in the applications. Therefore it is very necessary to make more research in these types of multivariate approximation which one has in the other sciences. In the following we describe some types of this kind.

2. H-SETS IN THE THREE-DIMENSIONAL DIRICHLET-PROBLEM

In simple cases, for instance for the Dirichlet-Problem of the potential equation in the plane, one has the approximation of the given boundary values along the boundary and this problem reduces to onedimensional approximation (many examples are computed by this way, compare f.i. Collatz[5]), but these cases will not be considered here corresponding to the title of this paper. But even the Dirichlet-Problem in the three-dimensional space requires a two-dimensional approximation along the boundary-surface.

EXAMPLE Let S the unit sphere in the rectangular x-y-z-space:

$$S = \{(x,y,z),\ x^2+y^2+z^2 < 1\} :$$

The function $u(x,y,z)$, for instance the temperature, should satisfy the stationary heat conduction equation

$$\Delta u = \frac{\partial^2 u}{\partial x^2} + \frac{\partial^2 u}{\partial y^2} + \frac{\partial^2 u}{\partial z^2} = 0 \quad \text{in} \quad S \tag{1.1}$$

and the boundary condition (given distribution of the temperature at the surface ∂S)

$$u = g(x,y,z) \quad \text{on} \quad \partial S \ . \tag{1.2}$$

Let be $w_\nu(x,y,z)$ for $\nu = 1,2,\ldots,p$ potential functions, with

$$\Delta w_\nu = 0 \quad \text{in} \quad S \quad (\nu = 1,\ldots,p) \tag{1.3}$$

(for instance $w_1 = 1$, $w_2 = x$, ...), then we have the linear Tschebyscheff-Approximation (abbreviated T.A.) (Meinardus[13]) of the given function g by the chosen functions w_ν. We have to determine constants $a_1,\ldots,a_p$ in such a way, that the error

$$\varepsilon(x,y,z) = \sum_{\nu=1}^{p} a_\nu w_\nu(x,y,z) - u(x,y,z) \tag{1.4}$$

along ∂S is as small as possible; from

$$|\varepsilon| \le \delta \quad \text{on} \quad \partial S \quad \text{follows} \quad |\varepsilon| \le \delta \quad \text{in} \quad S . \tag{1.5}$$

with δ = constant. So we get an exact inclusion theorem for the wanted solution u.

Numerical results Let us choose for illustration

$$g(x,y,z) = \sqrt{1+x^4+y^4+z^4}$$

and as functions w_ν polynomials which satisfy (1.3) and the symmetries of the problem:

$$w_1 = 1, \quad w_2 = (x^2+y^2+z^2)^2 - 5(x^2y^2+y^2z^2+z^2x^2), \quad w_3 = \ldots$$

We get for $p = 2$ (only two parameters a_1, a_2)

$$a_1 = 1.2619, \quad a_2 = 0.1557, \quad \delta = 0.0034,$$

that means the exact error bounds

$$|1.2619+0.1557w_2(x,y,z)-u(x,y,z)| \leq 0.0034 \quad \text{in} \quad S \ . \qquad (1.6)$$

Now we can apply the theory of H-sets (compare L. Collatz[4], Collatz-Krabs[9]) to see how good the approximation may be. The points P_1, P_2, P_3, P_4 (the unit-points on the coordinate axes and the point P_4 with $x = y = z = \sqrt{1/3}$, Figure 1) are an H-set; this H-set is not minimal, one could omit one of the points P_1, P_2, P_3 , but we have taken all these 3 points for reasons of symmetry).

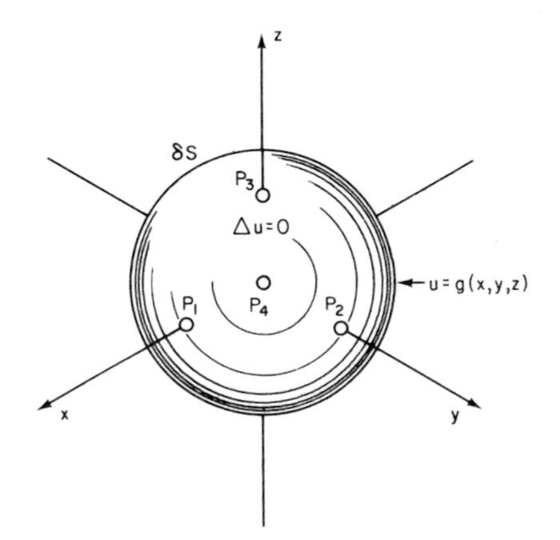

Figure 1

The errors are

$$w(P_j) - u(P_j) = 0.00339 \qquad (j = 1,2,3)$$
$$w(P_4) - u(P_4) = -0.00340$$

The minimal distance ρ therefore is

$$0.00339 \le \rho \le 0.00340$$

We have reached the best possible approximation with w_1, w_2 in the bounds of 4 decimals. If we wish to get higher accuracy, it is necessary to take other basic functions, for instance to add a function w_3 to the set w_1, w_2 .

3. LINEAR DIFFERENTIAL EQUATION

The Laplace equation $\Delta u = 0$ for a function $u(x,y)$ with two independent variables x,y in a given domain B has the convenient property, that one can use approximate solutions $w(x,y)$ which satisfy $\Delta w = 0$ (Wetterling[17], Whiteman[18]), but even in simple cases, for instance for variable coefficients in the differential equation, one has usually not solutions $w(x,y)$ with enough free parameters for approximating the boundary conditions along the boundary ∂B . Then one has to approximate the differential equation in the interior of B , this means a two-dimensional approximation problem and analogously three-dimensional approximation in the R^3 .

Let us consider the linear differential equation (with $x = x_1$, $y = y_2$)

$$Lu = \sum_{j,k=1}^{2} a_{jk} \frac{\partial^2 u}{\partial x_j \partial x_k} + \sum_{j=1}^{2} b_j \frac{\partial u}{\partial x_j} + cu = \phi \tag{3.1}$$

with given functions a_{jk}, b_j, c, ϕ , which depend on x_1, x_2 .

Then one can look for an approximate solution w of the differential equation:

$$u \simeq w = \sum_{j=1}^{p} \alpha_j w_j(x_1, x_2) \tag{3.2}$$

with chosen basic functions w_j with continuous derivatives up to the second order, for instance $w_1 = 1$, $w_2 = x_1$, $w_3 = x_2$, $w_4 = x_1^2$, ...

We can calculate the functions

$$h_j(x_1,x_2) = Lw_j(x_1,x_2) \tag{3.3}$$

and have to carry out the approximation with unknown constants α_j

$$\phi \simeq \sum_{j=1}^{p} \alpha_j h_j \qquad \text{or}$$

$$-\delta \le \sum_{j=1}^{p} \alpha_j h_j - \phi \le \delta \quad , \qquad \delta = \text{Min} \quad . \tag{3.4}$$

One can choose for simple boundary conditions such basic functions w_j , which satisfy these boundary conditions, but in more complicated cases one has to approximate also the boundary conditions (Simultaneous Approximation, Bredendiek[2], Bredendiek-Collatz[3]). Many open questions occur in this area. One has for linear Tschebyscheff approximation algorithms for testing whether a set of points, in which the modulus of the error $\varepsilon = w - u$ has local maxima, is an H-set (compare f.i. Collatz-Krabs[9]); but one knows H-sets in the multivariate case only for special classes of families of functions and one would wish to have available more general classes of H-sets. Again here much further research is necessary.

EXAMPLE We ask for a solution $u(x,y)$ of the differential equation:

$$Tu = -\Delta u + (2+x)u = 0 \quad \text{in} \quad B = \{x,y;\ |x|<1, |y|<1\} \tag{3.5}$$

and the boundary conditions

$$u = 0 \text{ on } \Gamma_1: x = -1, |y| \le 1 \text{ and } |x| < 1, y = \pm 1 \qquad (3.6)$$
$$u = 1 \text{ on } \Gamma_2: x = 1, |y| < 1, \text{ (Figure 2)}$$

We have singularities at the corners P_1, P_2 (Figure 2). $u(x,y)$ can be interpreted as distribution of temperature in a plate. We introduce in the usual way polar coordinates r_j, Φ_j at the corner P_j, compare Figure 2; we look for an approximate solution w of the form:

$$u \simeq w = \frac{2}{\pi} \sum_{j=1}^{2} \Phi_j + \sum_{\mu+2\nu \le k} a_{\mu\nu} x^{\mu} y^{2\nu} . \qquad (3.7)$$

The monotonicity principle allows the bounds for the error $\varepsilon = w - u$:

$$T\varepsilon \le 0 \text{ in } B, \ \varepsilon \le 0 \text{ on } \Gamma_1 \cup \Gamma_2 \text{ implies } \varepsilon \le 0 \text{ or } w \le u \text{ in } B . \qquad (3.8)$$

(One-sided Tschebyscheff Approximation, compare f.i. Collatz[6].) Analogously we can get an upper bound for u .

Combining these two principles of getting bounds we have the optimization problem for an upper and a lower bound w_1 and w_2 :

$$Tw_1 \ge 0, \ Tw_2 \le 0, \ w_1 - w_2 \le \gamma \text{ in } B, \ \gamma = \text{Min} \qquad (3.9)$$
$$w_1 \ge u, \ w_2 \le u \text{ on } \Gamma_1 \cup \Gamma_2 .$$

Then we have the inclusion

$$w_2 \le u \le w_1 \text{ in } B . \qquad (3.10)$$

We got the error bounds by this way:

k = Degree of polynomials	γ = Error bound for u
1	1
3	0.57
4	0.12
5	0.063
7	0.022

I thank Mr U. Grothkopf for numerical calculations on a computer.

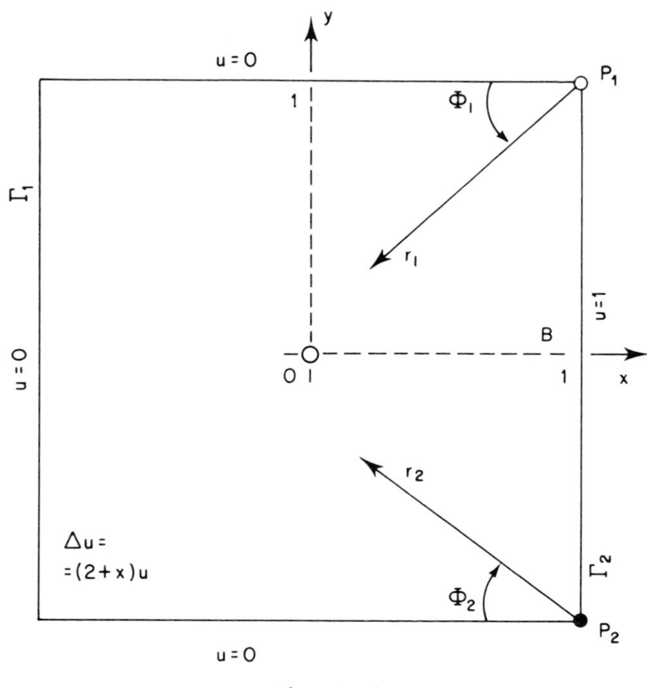

Figure 2

4. NONLINEAR DIFFERENTIAL EQUATIONS

If one has a nonlinear partial differential equation in a domain B (subset of the n-dimensional point space R_n)

$$Tu(x_1,\ldots,x_n) = 0 \quad \text{in} \quad B \tag{4.1}$$

with boundary conditions $Fu(x_1,\ldots,x_n) = 0$ on ∂B (given vector valued function F), then one has usually a multivariate approximation, which is nonlinear even if the approximation (3.2) is a linear one. One knows very little about this area. Even the question whether a set of extremal points is an H-set, is in the nonlinear case often very difficult. We illustrate this by an example.

EXAMPLE We consider the nonlinear boundary value problem for an unbounded domain $B = \{(x,y),\ 0 < x < \infty,\ |y| < 1\}$ (Figure 3):

$$Tu - \Delta u + u^2 = 0 \quad \text{in} \quad B \tag{4.2}$$
$$u = 0 \text{ for } 0 \le x < \infty,\ |y| = 1,\ u = 1 - y^2 \text{ for } x = 0,\ |y| \le 1 \quad .$$

We approximate $u(x,y)$ by functions $w(x,y,a)$ of the form

$$u \simeq w = (1-y^2)e^{-ax} \tag{4.3}$$

We get

$$Tw = \Phi_a e^{-ax} \quad \text{with} \quad \Phi_a = -2 + a^2(1-y^2) + (1-y^2)^2 e^{-ax} \quad .$$

The defect Tw has for $y = 0$, $a = 1.0825$ the graph of Figure 4 and we test with the aid of the theory of H-sets whether we have found the best possible approximation in the Tschebyscheff-sense for the defect zero with functions of the form (4.3).

We wish to decide whether the points $P = (0,0)$ and $Q = (s,0)$ (Figure 3) are an H-set. This is true, if (for two positive values a,b of the parameter) the inequalities

$\psi > 0$ for $x = 0$, $\psi < 0$ for $x = s$ with $\psi = \Phi_a e^{-ax} - \Phi_b e^{-bx}$

have no solution.

The first inequality $\psi > 0$ for $x = 0$ is equivalent to $a^2 > b^2$. After elementary considerations for the derivative $\partial\psi/\partial a$: we get that P and Q is an H-set, if $\zeta(a,s) = -2s-2a+a^2s+2se^{-as} < 0$. Figure 5 shows the graph of $\zeta(a,s) = 0$. We have $\zeta < 0$ for $0 < s < 1$, $0 < a < 2$. Therefore we have found the best approximation for functions $w(x,y,a)$ with $0 < a < 2$.

We can get also upper and lower bounds for the solution by using monotonicity principles and correspondingly one-sided Tschebyscheff approximation. For instance: $w(x,y)$ is for $a = 1$ an upper bound for u .

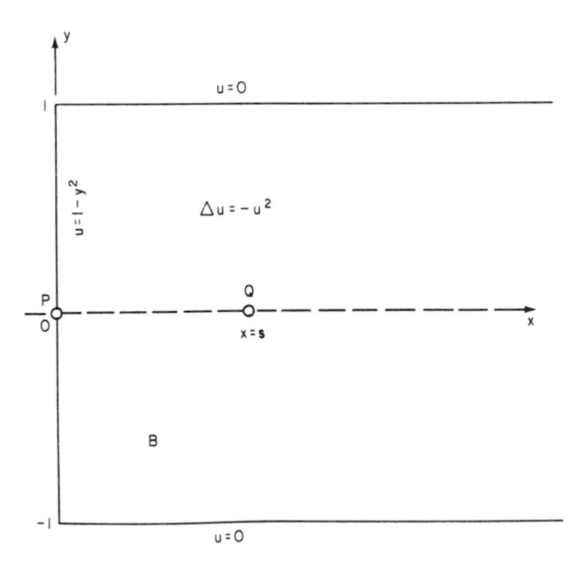

Figure 3

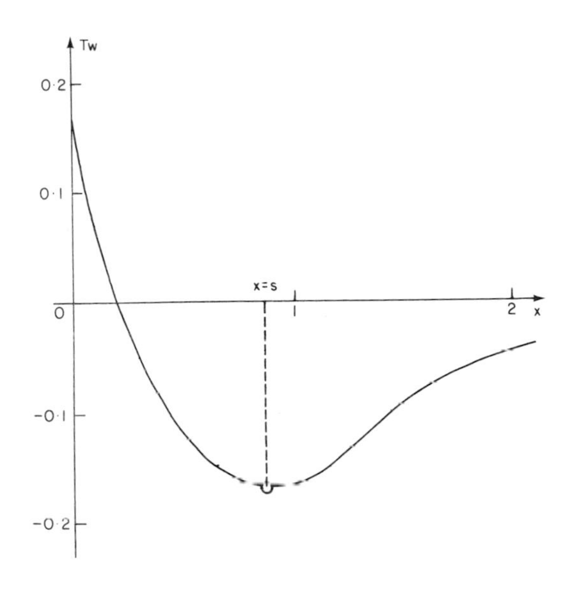

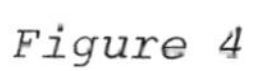
Figure 4

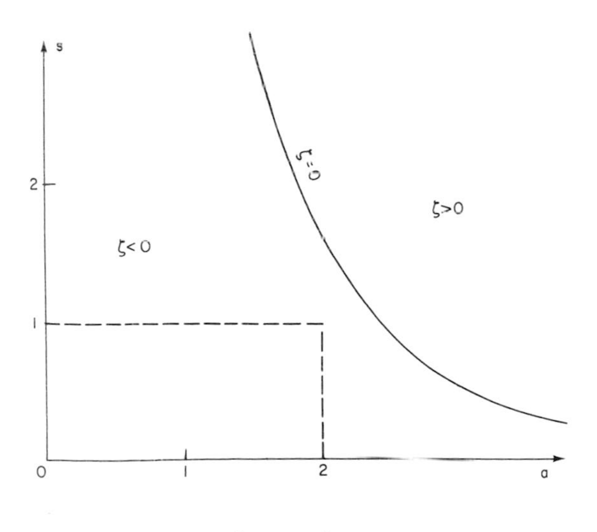

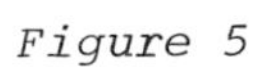
Figure 5

5. INTEGRAL EQUATIONS

Multivariate approximation occurs in integral equations even in the case that one asks for a solution $u(x)$ which depends only on one single variable. Consider the nonlinear Hammerstein equation

$$u(x) = \int_B K(x,t)\ \phi(u(t))\ dt + f(x) \quad , \tag{5.1}$$

where B is a measurable subset of Euclidean point space R^n; x means the vector $x = (x_1,\ldots,x_n)$. f and K are defined and continuous in B resp. in $B\times B$, and $\phi(s)$ is a continuous function defined in a suitable real interval (S_0,S_1) which may be unbounded.

There are several possibilities for inclusion of solutions, either using monotonicity properties or cone iteration (see Sprekels[15], Voss[16]) or approximation by degenerate kernels.

If the kernel $K(x,t)$ is degenerate, i.e. if $K(x,t)$ is of the form

$$K^*(x,t) = \sum_{j=1}^{p} v_j(x) w_j(t) \quad , \tag{5.2}$$

where v_j and w_j are continuous in B , then every solution of (5.1) is of the form

$$u(x) = f(x) + \sum_{j=1}^{p} c_j v_j(x) \quad ; \tag{5.3}$$

and (5.1) reduces to a nonlinear system of p equations for the p constants c_j .

If the kernel $K(x,t)$ is not degenerate, we try to approximate $K(x,t)$ by a degenerate kernel $K^*(x,t)$. We can do this in two different ways:

A) One takes for $K^*(x,t)$ the form

$$K^*(x,t) = \sum_{j=1}^{p} v_j(x,a_{j\nu})w_j(t,b_{j\mu}) \quad , \tag{5.4}$$

where $v_j(x,a_{j\nu})$ is a chosen function of x and some parameters $a_{j1},a_{j2},\ldots$, and analogously $w_j(t,b_{j\mu})$. One tries to determine the parameters $a_{j\nu},b_{j\mu}$ in such a way that

$$|K^*(x,t) - K(x,t)| \le \delta \quad \text{for} \quad (x,t) \in BxB \ , \quad \delta = \text{Min} \quad . \tag{5.5}$$

Even for the simplest case that B is a real interval $J = |a,b|$, we have a two-dimensional Tschebyscheff approximation problem; if B is a subset of the x_1-x_2-plane, one has a four-dimensional Tschebyscheff approximation problem.

B) Given p one asks for the best choice of the functions $v_j(x),w_j(t)$ in (5.2) such that (5.5) is satisfied with δ as small as possible. Now there are not finitely many parameters $a_{j\nu},b_{j\mu}$ to determine but functions. This is a problem of approximation of functions of several variables by functions of fewer variables (compare Colomb[11], Sprecher[14], Collatz[1], Flügge[10]), in this case a problem of "Product-sum-type" (Collatz[1], p.17). Very little is known about Tschebyscheff approximation of this kind.

EXAMPLE The kernel

$$K(x,t) = \exp(-\sqrt{1+x^2+t^2-x^2t^2}) \quad \text{in } JxJ = (x,t), |x|\le 1, |t|\le 1 \tag{5.6}$$

should be approximated by the degenerate kernel

$$K^*(x,t) = e^{-a_1-a_2(x^2+t^2)} = v(x)v(t)$$
$$\text{with} \quad v(z) = \exp|-\frac{a_1}{2} - a_2 z^2| \quad . \tag{5.7}$$

For $a_1 = 1.0781$, $a_2 = 0.2284$ one gets

$$|K^*(x,t) - K(x,t)| \leq 0.0277 \quad \text{in} \quad J x J \quad .$$

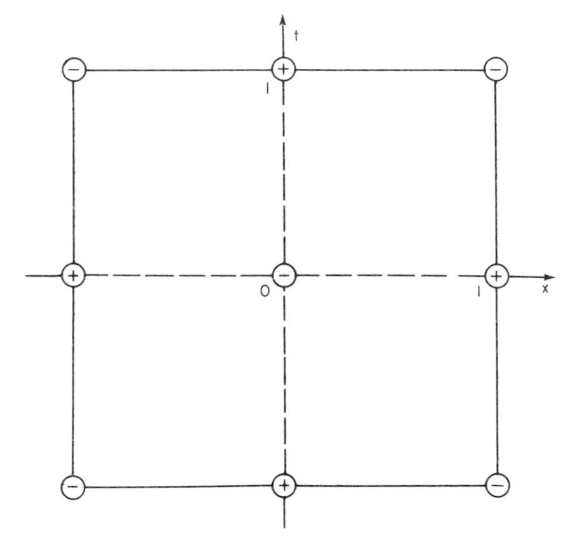

Figure 6

Figure 6 shows the distribution of the extremal points, in which the modulus of the error K^*-K takes its maximum; in these nine points it is indicated by + or − , whether the error K^*-K is positive or negative. The points (0,0)(0,1)(1,1) are an H-set. One cannot expect in this example better results with the kernel K^* (which contains only two parameters) because the kernel $K(x,t)$ is constant $= e^{-\sqrt{2}}$ along the whole boundary $|x| = 1$ and $|t| = 1$.

6. OTHER PROBLEMS

One could add many other types of problems with multivariable approximation. A lot of linear and nonlinear examples are treated on the computer (compare for instance Collatz[5] also for differential equation of higher order (biharmonic plate-equation)).

One has had success also in boundary value problems which seemed to be inaccessible even some years ago, for instance some types of free boundary value problems. We mention only the classical Stefan problem (compare Baiocchi[1], Hoffmann[12]): We consider a melting ice-cover of a lake in the halfplane $x > 0$ of an x-y-plane. At time t a strip $0 < x < s(t)$ of free water is at temperature $u = u(x,t) > 0$; the part $x > s(t)$ is covered with ice. We have for the unknown function u in the domain $B = \{(x,t),\ 0 < t,\ 0 < x < s(t)\}$ (Figure 7) the differential equation

$$Lu = \frac{\partial u}{\partial t} - \frac{\partial^2 u}{\partial x^2} = 0 \quad \text{in} \quad B \tag{6.1}$$

and the boundary conditions

$$\frac{\partial u}{\partial x} = - g(t) \quad \text{on} \quad \Gamma_1 \quad (x = 0,\ t > 0) \tag{6.2}$$

$$u = f(x) \quad \text{on} \quad \Gamma_2 \quad (t = 0,\ 0 < x < s(0))$$

$$x = s(t),\ u = 0,\ \frac{\partial u}{\partial x} = - \frac{ds}{dt} \quad \text{on} \quad \Gamma_3 \quad (x > s(0),\ t > 0)\ .$$

One wants to determine the free boundary $x = s(t)$.

It was possible with the aid of a nonlinear integral equation, with approximation methods and Schauder's fixed point theorem to get exact inclusions for the free boundary $x = s(t)$ in certain cases. (A detailed numerical example is given in Collatz[8].)

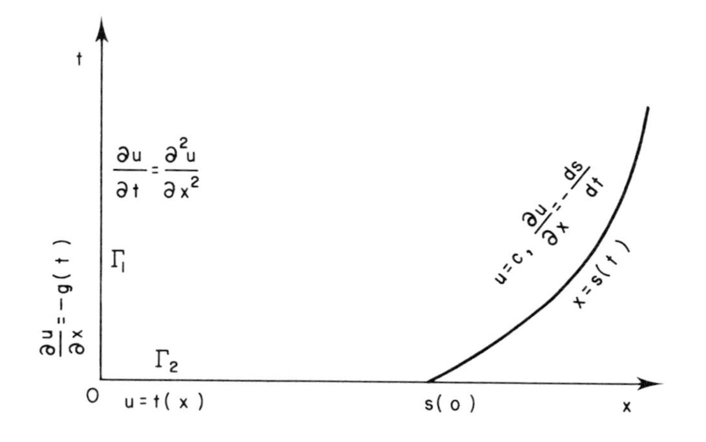

Figure 7

REFERENCES

1. Baiocchi, C. (1974). "Free Boundary Problems in the Theory of Fluid Flow Through Porous Media", Proc. Intern. Congress of Math., Vol.2, Vancouver.

2. Bredendiek, E. (1970). Charakterisierung und Eindeutigkeit bei simultanen Approximationen, *Z. angew. Math. Mech.* 50, 403-410.

3. Bredendiek, E. and Collatz, L. (1976). Simultan-Approximation bei Randwertaufgaben, *Intern. Ser. Num. Math.* 30, 147-174.

4. Collatz, L. (1956). Approximation von Funktionen bei einer und bei mehreren unabhängigen Veränderlichen, *Z. angew. Math. Mech.* 36, 198-211.

5. Collatz, L. (1965). Tschebyscheffsche Approximation, Randwertaufgaben und Optimierungsaufgaben, *Wiss. Zeitschr. d. Hochschule für Architektur und Bauwesen, Weimar* 12, 504-509.

6. Collatz, L. (1970). "Einseitige Tschebyscheff-Approximation bei Randwertaufgaben", Proc. Internat. Conference on Constructive Funct. Theory, 151-162, Varna, Bulgaria.

7. Collatz, L. (1972). "Approximation by Functions of Fewer Variables", Proc. Conference Theory Ordin. Part. Diff. Equ., Dundee 1972, Lect. Notes Math. 280, 16-31, Springer.

8. Collatz, L. (to appear). "The Numerical Treatment of Some Singular Boundary Value Problems", Proc. Conference Num. Methods, Dundee 1977.

9. Collatz, L. and Krabs, W. (1973). "Approximations Theorie, Tschebyscheffsche Approximation mit Anwendungen", 208p, Teubner, Stuttgart.

10. Flügge, W. (1975). "Approximation durch Funktionen von wenigen Veränderlichen", Diss. Univ. Hamburg.

11. Golomb, U. (1959). Approximation by Functions of Fewer Variables, Part B, *in* "On Numerical Approximation" (Ed. R.E. Langer), 311-327, Madison.

12. Hoffmann, K.H. (1977)."Lecture on Free Boundary Value Problems", Conference Num. Solut. Diff. Equ. Oberwolfach, May 1977, to appear in *Internat. Ser. Num. Math.* 39.

13. Meinardus, G. (1967). "Approximation of Functions, Theory and Numerical Methods", 198p, Springer.

14. Sprecher, D.A. (1968). On Best Approximation of Functions of Two Variables, *Duke Math. Journ.* 35, 391-397.

15. Sprekels, J. (to appear). Iterationsverfahren zur Einschließung positiver Lösungen superlinearer Integralgleichungen, *ISNM*.

16. Voss, H. (to appear). Existence and Bounds for Positive Solutions of Superlinear Uryson Equations, *Applicable Analysis*.

17. Wetterling, W. (to appear). Lecture on Error Bounds for Solutions of Singular Elliptic Boundary Value Problems at Mathem. Forschungsinstitut Oberwolfach, *Internat. Ser. Num. Math.*

18. Whiteman, J.R. (to appear). Lecture about Finite Element Methods for Singular Elliptic Boundary Value Problems at Mathem. Forschungs-Institut Oberwolfach, 19 May 1976, *Internat. Ser. Num. Math.*

MULTIVARIATE APPROXIMANTS WITH BRANCH CUTS

J.S.R. Chisholm

Mathematical Institute
University of Kent
Canterbury, England

1. INTRODUCTION

In 1973, I proposed a two-variable generalisation of diagonal Padé approximants[1]; these approximants have been referred to as "Chisholm approximants" by various authors. A group of us in the University of Kent generalised these rational polynomial approximants to N variables, defined off-diagonal approximants and a "rotationally covariant" variant, studied analytic and algebraic properties, wrote programmes, and investigated the numerical accuracy of the various approximants. This work has been reviewed on three occasions[2,3,4], and these reviews give a complete list of references for these N-variable rational approximants.

The numerical investigations showed that, as expected, the two-variable rational approximants converged to functions represented by the defining power series inside the polycylinder of convergence, provided an appropriate analytic continuation beyond the polycylinder, and represented the singularities of the approximants. However, convergence was not fast when the defining series represented functions with branch points; a rational approximant is single-valued, and so is not especially suitable for representing well one

Riemann sheet of a multivalued function, which normally possess discontinuities along cuts in each complex plane.

Shafer[5] has defined a one-variable generalisation of Padé approximents, "quadratic approximants", which are two-valued complex functions. These approximants represent multi-valued functions more accurately than Padé approximants do, and Leslie Short has shown[6] that they are more effective than Padé approximants for calculating elementary particle theory matrix elements, which are multi-valued functions. Since a wide variety of functions arising in mathematical physics, fluid mechanics and other subjects are multi-valued functions of several parameters which can be expressed as multivariate power series, there is good reason to look for a multivariate generalisation of Shafer's approximants. Since Easter, I have defined diagonal[7] and off-diagonal[8] multivariate "t-power approximants"; these are functions defined on t Riemann sheets. This work has been reported[9] at the recent Dundee Conference on Numerical Analysis; for this reason, I shall only give a brief account of these new approximants. I shall report some numerical results worked out by Leslie Short, and I shall discuss the "general philosophy of approximation" which is suggested by the wide range of multivariate approximants now defined.

2. N-VARIABLE t-POWER APPROXIMANTS

First, I shall define two-variable off-diagonal t-power approximants; then I shall comment on the N-variable generalisation and on diagonal approximants.

Two-variable approximants are defined from a formal double power series

$$f(\underset{\sim}{z}) = \sum_{\underset{\sim}{\gamma}=\underset{\sim}{0}}^{\underset{\sim}{\infty}} c_{\underset{\sim}{\gamma}} \underset{\sim}{z}^{\underset{\sim}{\gamma}} \tag{2.1}$$

where $\underset{\sim}{z} = (z_1, z_2)$, $\underset{\sim}{\gamma} = (\gamma_1, \gamma_2)$ and

$$\underset{\sim}{z}^{\underset{\sim}{\gamma}} \equiv z_1^{\gamma_1} z_2^{\gamma_2} .$$

We proceed to define $(t+1)$ polynomials

$$p^{(k)}(\underset{\sim}{z}) \equiv \sum_{\underset{\sim}{\alpha} \subset S_k} p_{\underset{\sim}{\alpha}}^{(k)} \underset{\sim}{z}^{\underset{\sim}{\alpha}} \quad (k = 0,1,\ldots,t) , \tag{2.2}$$

with coefficients $\{p_{\underset{\sim}{\alpha}}^{(k)}\}$ defined over $k+1$ integer lattice rectangles

$$S_k = \{\underset{\sim}{\alpha} | 0 \le \alpha_i \le m_{i,k} \quad ; \quad i = 1,2\} . \tag{2.3}$$

The formal k^{th} power of the series (2.1) is written

$$f^k(\underset{\sim}{z}) \equiv \sum_{\underset{\sim}{\alpha}=\underset{\sim}{0}}^{\underset{\sim}{\infty}} c_{\underset{\sim}{\gamma}}^{(k)} \underset{\sim}{z}^{\underset{\sim}{\alpha}} , \tag{2.4}$$

and the coefficient of $\underset{\sim}{z}^{\underset{\sim}{\varepsilon}}$ in the formal series

$$E(\underset{\sim}{z}) = \sum_{k=1}^{t} p^{(k)}(\underset{\sim}{z}) f^k(\underset{\sim}{z}) + p^{(0)}(\underset{\sim}{z}) \tag{2.5}$$

is then

$$c_{\underset{\sim}{\varepsilon}} = \sum_{k=1}^{t} \left[\sum_{\underset{\sim}{\alpha} \subset S_k \cap S_{\underset{\sim}{\varepsilon}}} p_{\underset{\sim}{\alpha}}^{(k)} c_{\underset{\sim}{\varepsilon}-\underset{\sim}{\alpha}}^{(k)} \right] + p_{\underset{\sim}{\varepsilon}}^{(0)} \delta(\underset{\sim}{\varepsilon} \subset S_0) , \tag{2.6}$$

where $S_{\underset{\sim}{\varepsilon}}$ is the integer lattice rectangle

$$S_k = \{\underset{\sim}{\alpha} | 0 \le \alpha_i \le \varepsilon_i \quad ; \quad i = 1,2\} \tag{2.7}$$

and

$$\delta(\underset{\sim}{\varepsilon} \subset S_0) = \left\{ \begin{array}{lll} 1 & , & \underset{\sim}{\varepsilon} \subset S_0 \\ 0 & , & \underset{\sim}{\varepsilon} \not\subset S_0 \end{array} \right\} . \tag{2.8}$$

The coefficients $\{p_{\underset{\sim}{\alpha}}^{(k)}\}$ are determined by equating to zero certain $e_{\underset{\sim}{\varepsilon}}$ and linear combinations of $e_{\underset{\sim}{\varepsilon}}$, which I shall describe. Normally, these equations determine the polynomials $p^{(k)}(\underset{\sim}{z})$ uniquely; the t-power approximant $f(\underset{\sim}{z};S_\ell)$ is then defined as the solution of the equation

$$\sum_{k=1}^{t} p^{(k)}(\underset{\sim}{z})\,[f(\underset{\sim}{z};S_\ell)]^k + p^{(0)}(\underset{\sim}{z}) = 0 \quad . \tag{2.9}$$

The solution is defined on t Riemann sheets; the Riemann sheet corresponding to the series (2.1) is that for which

$$f(\underset{\sim}{0};S_\ell) = c_{\underset{\sim}{0}} \quad . \tag{2.10}$$

The definition of the original approximants[1] was chosen to ensure satisfaction of several specific properties:

(i) symmetry between the variables;

(ii) the projection property, that an approximant with one variable zero is the Padé approximant of the series with that variable zero;

(iii) reciprocal covariance, ensuring that an approximant to the reciprocal series is the reciprocal of an approximant to the original series;

(iv) homographic covariance, which is closely associated with the analytic continuation properties of both diagonal Padé approximants and Chisholm approximants.

Hughes Jones and Makinson[10] studied the full algebraic structure of the linear equations I had proposed. When the linear equations corresponding to points on the integer lattice are written down in sets, each set corresponding to points on an L-shaped "prong" on the lattice, the matrix of the set of equations is block lower diagonal, with square blocks on the diagonal. This "prong structure" ensures good algebraic properties of the system of equations, and enables the system to be solved without enormous difficulty.

The system of equations which I have proposed[7,8] is based upon the prong method, although I have also suggested some variants. In particular, I have suggested replacing some equations linear in $\{e_{\underset{\sim}{\varepsilon}}\}$ by $(t-1)$ other equations; these ensure that (2.9) is satisfied when $\underset{\sim}{z} = \underset{\sim}{0}$ and $f(\underset{\sim}{0};S_\ell)$ is given the values of the function at the origin on $(t-1)$ other Riemann sheets. Since we may not know function values at these other points, this "multi-point" variant may not be practicable.

The use of the prong method automatically ensures that the system of equations has good algebraic and computational properties. Since the prong structure is closely associated with properties (i)-(iv) set out above, these properties are, with a few minor qualifications, satisfied. The properties are fully satisfied for two-variable and three-variable approximants by the scheme which fits the function at the origin on $(t-1)$ other sheets. The symmetry property cannot be fully satisfied for $N \geq 4$ variables. Numerical experiment has shown, however, that the minor deviations from properties (i)-(iv) are not likely to affect numerical accuracy very much.

3. NUMERICAL RESULTS

Leslie Short has written a programme to compute quadratic $(t = 2)$ two-variable approximants, and has used it to evaluate "diagonal" approximants to four multi-valued two-variable functions from their power series. He has compared their accuracy with that of Chisholm rational approximants. In Tables 1-4, the moduli of the errors of quadratic approximants (Q.A.) and rational approximants (R.A.) are given, for a range of real values of the variables $z_1 = x$ and $z_2 = y$; the function values are of order unity unless they are zero. When one variable is zero, the approximant is very similar to a Shafer one-variable approximant. One general point emerges; the quadratic approximants approximate, often quite accurately, well beyond the radius of convergence of the series (2.1), even at points which would normally be placed on the "cut"; at such points, there is no hope that a single-valued rational approximant can represent the discontinuity across the cut. I shall now discuss the results given in the Tables; an error written as aEb is $a.10^b$, and an asterisk indicates the common error of two complex conjugate approximants on a "cut".

Table I Moduli of Errors in $f(x,y) = 1+\frac{1}{2}x+\ln(1-x-y)$

y	x	[6,6/6,6]R.A.	[4,4/4,4/4,4]Q.A.	[7,7/7,7/7,7]Q.A.
-4	-4	1E0	1E-4	1E-8
	-2	5E-1	1E-5	3E-8
	0	2E-5	2E-6	0
	2	2E-2	1E-4	4E-6
	4	7E-2	9E-2	4E-2
-2	-4	2E-1	3E-5	1E-8
	-2	3E-3	2E-6	0
	0	2E-7	5E-8	0
	2	6E-4	2E-4	0
	4		1E-2*	7E-4*

*Equal errors of two complex values

Table I Cont'd.

0	-4	2E-5	4E-6	2E-10
	-2	3E-7	6E-8	0
	0	0	0	0
	2		2E-3*	6E-6*
	4		1E-2*	3E-5*
2	-4	5E-2	2E-4	4E-6
	-2	1E-3	4E-6	2E-7
	0		2E-3*	6E-6*
	2		5E-3*	1E-6*
	4		2E-2*	5E-6*
4	-4	Not Given	4E-3	6E-4
	-2		2E-3*	8E-4*
	0		2E-3*	2E-5*
	2		2E-2*	3E-5*
	4		4E-2*	3E-4*

The number of terms of (2.1) used to calculate the [6,6/6,6] R.A. and the [4,4/4,4/4,4] Q.A. are approximately equal, so that it is reasonable to compare their accuracy. The function $f(x,y)$ is infinitely many-valued, but its series expansion corresponds to a single Riemann sheet. At nearly every point, the Q.A. is one to four orders of magnitude more accurate than the R.A. At points "on the cut", where $1-x-y < 0$, the R.A. is necessarily inaccurate and its error

*Equal errors of two complex values

is not given. The introduction of only *one* extra Riemann sheet produces a dramatic improvement in the representation of an infinite-sheeted function. The [7,7/7,7/7,7] Q.A. gives a very accurate representation of the function, sometimes accurate to order 10^{-10} or better.

Table II Moduli of Errors in $f(x,y) = x+(1-x-y)^{-1/3}$

y	x	[8,8/8,8] R.A.		[6,6/6,6/6,6] Q.A.	
-4	-4	2E0		2E-3	
	-2	3E-5		4E-6	
	0	5E-8		1E-10	
	2	2E-5		3E-3	
	4	8E-3		5E0	
-2	-4	1E-4		3E-5	
	-2	2E-6		2E-7	
	0	2E-10		0	
	2	1E-5		9E-3	
	4	5E-2	(real root)	5E-2*	(complex roots)
0	-4	1E-7		1E-9	
	-2	1E-9		0	
	0	0		0	
	2	6E0	(real root)	4E-6*	
	4	1E2	(real root)	3E-5*	
2	-4	4E-4		1E-3	
	-2	2E-5		5E-4	

*Equal errors of two complex values

Table II Cont'd.

	0	0	(real root)	3E-6*
	2	2E0	(real root)	3E-5*
	4	5E-1	(real root)	4E-4*
4	-4	9E-2		1E-2
	-2	1E0	(real root)	9E-3*
	0	1E0	(real root)	7E-6*
	2	6E-1	(real root)	1E-4*
	4	1E0	(real root)	1E-2*

Again, the [6,6/6,6/6,6] Q.A. is generally more accurate than the [8,8/8,8] R.A. , but at some points such as $y = -4$, $x = 2$ or 4 , the R.A. is more accurate. The function $f(x,y)$ is defined on three Riemann sheets. On the "cut" the two values of the Q.A. give a fairly accurate representation of the complex-valued roots. The values of the (real) R.A. are also given on the cut, and generally do not approximate any value of the function. However, at $y = -2$, $x = 4$, and especially at $y = 2$, $x = 0$, the R.A. approximates the real root while the Q.A. approximates the two complex roots. So from the power series on one Riemann sheet, we are obtaining approximations on all three Riemann sheets. This is a very strong form of analytic continuation, and offers a formidable challenge to analysts!

---

*Equal errors of two complex values

Table III Moduli of Errors in $f(x,y) = (2-y)^{-1}\ln(1-x-y)$

y	x	[8,8/8,8] R.A.	[5,5/5,5/5,5] Q.A.
-4	-4	1E-2	3E-3
	-2	2E-3	2E-4
	0	5E-8	5E-9
	2	4E-2	1E-3
	4	1E-8	1E-10
-2	-4	1E-3	4E-5
	-2	2E-4	8E-6
	0	3E-10	0
	2	1E-9	0
	4		2E-2*
0	-4	2E-7	3E-8
	-2	5E-10	0
	0	0	0
	2		2E-3*
	4		8E-3*
4	-4	1E-6	1E-10
	-2		3E-2*
	0		3E-3*
	2		2E-1*
	4		2E0*

*Equal errors of two complex values

This function has a pole surface as well as a surface of infinitely-sheeted branch points. Both the [8,8/8,8] R.A. and the [5,5/5,5/5,5] Q.A. approximate the function well when $1-x-y \geq 0$, but the Q.A. is consistently better. On the cut $1-x-y < 0$, the Q.A. approximates two complex values with moderate accuracy.

Table IV Moduli of Errors in $f(x,y) = (1-x)^{-1} + (1-x-y)^{-1/3}$

y	x	[8,8/8,8] R.A.	[5,5/5,5/5,5] Q.A.
-2	-2	7E-2	3E-2
	0	1E-9	0
	2	2E0	6E0
0	-2	0	0
	0	0	0
	2		4E-5*
2	-2	2E1	2E-1
	0		2E-5*
	2		2E-2*
0	-4	1E-7	1E-10
0	4		4E-7*
-4	0	1E-7	1E-9
4	0		3E-5*

*Equal errors of two complex values

This function also has surfaces of both poles and branch points; it is three-valued. The Q.A.'s represent the function reasonably well at the points shown, many of which have $x = 0$ or $y = 0$. These include several points "on the cut", where the R.A.'s fail. This example is a warning not to expect too much of a method, especially when applied in awkward circumstances. Had we used a suitable cubic approximant, we would have obtained the exact result, on all three Riemann sheets!

4. DISCUSSION

It is clear that the quadratic two-variable approximants, used with care, are a useful method of calculating functions of two variables with branch points. The range of approximants now defined, diagonal and off-diagonal, with arbitrary power t and number of variables N, is vast. But the same basic concepts extend much further, becoming essentially a "philosophy of approximation". There is no reason why the basic function (2.5) should be a polynomial in f; we could, for example, use the formal exponential $\exp f$ of the given series to define

$$E(\underset{\sim}{z}) = p^{(2)}(\underset{\sim}{z}) \exp f(\underset{\sim}{z}) + p^{(1)}(\underset{\sim}{z})\, f(\underset{\sim}{z}) + p^{(0)}(z) . \qquad (4.1)$$

The coefficients in $p^{(2)}, p^{(1)}, p^{(0)}$ could be defined just as before, and the approximant $f(\underset{\sim}{z};S_\ell)$ found by solving the transcendental equation

$$p^{(2)}(\underset{\sim}{z}) \exp f(\underset{\sim}{z};S_\ell) + p^{(1)}(\underset{\sim}{z})\, f(\underset{\sim}{z};S_\ell) + p^{(0)}(\underset{\sim}{z}) = 0 . \qquad (4.2)$$

This is no harder than solving, say, a quintic equation. An associated idea is to operate on the given series f initially with some suitable invertible operation O, form an approximant, and then operate with 0^{-1} . If R,Q denote the formation of rational and quadratic approximants respectively, approximants of this type are

$$0^{-1}R0f \quad , \quad 0^{-1}Q0f \quad .$$

There is a wide possible range of operations 0 . Two simple examples are $0 = \ln$, $0^{-1} = \exp$ and $0 =$ integration , $0^{-1} =$ differentiation, with respect to some parameter. This enormous range of possibilities was foreshadowed[11] by Baker's "D log method", and was exemplified recently[12] by Fisher's two variable "partial differential approximants". The underlying philosophy can be summarised:

a. Given a power series f , an approximant is constructed whose power series expansion matches as many terms as possible of the given series.

b. The construction will involve the determination from a set of linear equations of the coefficients of a set of polynomials. This set of equations should be constructed by the prong method[10,7,8], perhaps incorporating some multi-point equations. This will ensure good algebraic and numerical properties, and the satisfaction or near-satisfaction of the properties of symmetry and projection, together with appropriate covariance properties.

c. The defining expression $F(z)$ should be chosen so that the analytic structure of the approximant matches those of the unknown function as closely as possible. In view of the complexity of functions of even two complex variables, and the wide variety of approximants that this "philosophy" allows, the choice of approximant becomes an art.

We have talked about the analytic structure of functions of several variables. Defining singularities of multivariate functions is a difficult matter, and it may eventually turn out that the best way of classifying multivariate functions is through the approximation methods discussed here. However, convergence theorems are not going to be easy to establish.

REFERENCES

1. Chisholm, J.S.R. (1973). Rational Approximants Defined from Double Power Series, *Math. Comp.* 27, 124.
2. Chisholm, J.S.R. (1974). "Rational Polynomial Approximants Defined from Power Series in N Variables", Proc. 3rd Marseilles Conference on Advanced Computational Methods in Theoretical Physics (Ed. A. Visconti, C.N.R.S. Marseilles).
3. Chisholm, J.S.R. (1976). Rational Polynomial Approximants in N Variables, *in* "The Padé Approximants Method and its Applications to Mechanics" (Ed. H. Cabannes) Springer-Verlag.
4. Chisholm, J.S.R. (1977). "N-variable Rational Approximants", Proceedings of 1976 Tampa Conference on Rational Approximation (Ed. E.B. Saff) Academic Press.
5. Shafer, R.E. (1974). On Quadratic Approximation, *S.I.A.M. Jour. Num. An.* 11, 447.
6. Chisholm, J.S.R. and Short, L. "Moving Branch Cuts in Feynman Integrals using Padé Approximants and their Generalisations", Proc. 1977 St. Maximin Conference on Advanced Computational Methods in Theoretical Physics (Ed. A. Visconti, C.N.R.S. Marseilles, to be published).
7. Chisholm, J.S.R. (1977). Multivariate Approximants with Branch Points I, *Proc. Roy. Soc.* 358, 351-366.
8. Chisholm, J.S.R. Multivariate Approximants with Branch Points II, *Proc. Roy. Soc.* (submitted for publication).
9. Chisholm, J.S.R. Multivariate Approximants with Branch Points, Proc. 1977 Dundee Conf. Num. Analysis (Ed. A. Watson) (to be published) Springer-Verlag.
10. Hughes Jones, R. and Makinson, G.J. (1974). The Generalisation of Chisholm Rational Polynomial Approximants to Power Series in Two Variables, *J.I.M.A.* 13, 299.
11. Baker, G.A. Jr. (1961). Application of the Padé Approximation Method to Investigation of some Magnetic Properties of the Ising Model, *Phys. Rev.* 124, 768.

12. Fisher, M.E. (1977). Series Expansion Approximants for Singular Functions of Many Variables, *in* "Statistical Mechanics and Statistical Methods in Theory and Application" (Ed. U. Landman) Plenum.

SERENDIPITY-TYPE BIVARIATE INTERPOLATION

F.-J. Delvos[†], H. Posdorf[††], W. Schempp[†]

[†]*Gesamthochschule Siegen*
Siegen, Germany

[††]*Rechenzentrum Universität Bochum*
Bochum, Germany

1. INTRODUCTION

Serendipity interpolation has been developed by the engineers[12] to obtain rectangular finite elements which are more useful than Lagrangian elements. Gordon and Hall[5] were the first workers to use blending interpolation in finite element construction. More recently Lancaster and Watkins[8,9,10,11] applied blending interpolation methods to the construction of other types of finite elements.

This paper has two objectives. One is to demonstrate the use of Boolean interpolation in the construction of serendipity elements. The other is to present new serendipity-type elements which are not covered by the approach of Gordon and Hall.

2. THE SERENDIPITY PROJECTOR

Let $S = [0,h]\times[0,h]$ be a square and let $C(S)$ denote the space of continuous functions on S. We consider three partitions of the interval $[0,h]$:

$$\{x_i\} : 0 = x_0 < x_1 < \ldots < x_m = h \quad ,$$

$$\{\bar{x}_i\} : 0 = \bar{x}_0 < \bar{x}_1 < \ldots < \bar{x}_{\bar{m}} = h \quad ,$$

$$\{\bar{\bar{x}}_i\} : 0 = \bar{\bar{x}}_0 < \bar{\bar{x}}_1 < \ldots < \bar{\bar{x}}_{\bar{\bar{m}}} = h \quad .$$

Using univariate Lagrange interpolation we construct *parametric projectors* P_s and P_t associated with $\{x_i\}$:

$$P_s(f) = \sum_{i=0}^{m} f(x_i,t)L_{i,m}(s) \ , \quad P_t(f) = \sum_{j=0}^{m} f(s,x_j)L_{j,m}(t) \ ,$$

where the functions $L_{i,m}$, $L_{j,m}$ denote the dual (shape) functions of univariate Lagrange interpolation. The parametric projectors $\bar{P}_s$, $\bar{P}_t$ associated with $\{\bar{x}_i\}$ as well as $\bar{\bar{P}}_s$, $\bar{\bar{P}}_t$ associated with $\{\bar{\bar{x}}_i\}$ are defined similarly.

We assume that the *containment relations* $\{x_i\} \subset \{\bar{x}_i\} \subset \{\bar{\bar{x}}_i\}$ are satisfied. Then the six parametric projectors $P_s,\ldots,\bar{\bar{P}}_t$ commute. It follows from a general result of Gordon[3] that the parametric projectors generate a *distributive lattice* $L = L(P_s,P_t,\bar{P}_s,\bar{P}_t,\bar{\bar{P}}_s,\bar{\bar{P}}_t)$. We consider special projectors of L .

The *product projector*

$$P_s\bar{P}_t(f) = \sum_{i=0}^{m} \sum_{j=0}^{\bar{m}} f(x_i,\bar{x}_j)L_{i,m}(s)L_{j,\bar{m}}(t)$$

defines *tensor product Lagrange interpolation.*

The *Boolean sum projector*

$$P_s \oplus P_t = P_s + P_t - P_sP_t$$

defines *blending Lagrange interpolation* (Gordon[4]).

The projector

$$P_s\bar{P}_t \oplus \bar{P}_sP_t = P_s\bar{P}_t + \bar{P}_sP_t - P_sP_t$$

defines *discrete blending interpolation* (Gordon-Hall[5]).

We will now consider the projector

$$B = P_s\bar{\bar{P}}_t \oplus \bar{P}_s\bar{P}_t \oplus \bar{\bar{P}}_sP_t$$

which we call *serendipity projector*. It follows from the containment relations that the following representation of the serendipity projector B is valid:

$$B = P_s\bar{\bar{P}}_t + \bar{P}_s\bar{P}_t + \bar{\bar{P}}_sP_t - P_s\bar{P}_t - \bar{P}_sP_t \quad .$$

This representation in terms of product projectors is useful for the determination of the dual functions of serendipity interpolation. For the interpolation properties of the serendipity projector B we will use the notation of precision set of an interpolation projector $P \subset L$.

The *precision set* $S(P)$ of $P \subset L$ is the set of interpolation points, i.e. we have: $(x,y) \in S(P)$ iff $P(f)(x,y) = f(x,y)$ for all $f \in C(S)$. For example, the precision set of the parametric projectors are lines:

$$S(P_s) = \{(x_i,t)\} \ , \ S(P_t) = \{(s,x_i)\} \ , \ \ldots ,$$

and the precision sets of the product projectors are rectangular grids:

$$S(P_sP_t) = \{(x_i,x_j)\} \ , \ S(P_s\bar{P}_t) = \{(x_i,\bar{x}_j)\} \ , \ \ldots \quad .$$

Gordon and Wixom[6] pointed out that

$$L := \{S(P) \ : \ P \in L\}$$

is a lattice of point sets and they stated the following.

THEOREM 1 (Precision set isomorphism) The mapping $P \leftrightarrow S(P)$

is a lattice isomorphism such that

$$PQ \leftrightarrow \mathcal{S}(P) \cap \mathcal{S}(Q) \quad ,$$

$$P \oplus Q \leftrightarrow \mathcal{S}(P) \cup \mathcal{S}(Q) \quad .$$

An application of the isomorphism theorem yields

THEOREM 2 The precision set of the serendipity projector B is the union of three grids:

$$\mathcal{S}(B) = \{(x_i,\bar{\bar{x}}_j)\} \cup \{(\bar{x}_i,\bar{x}_j)\} \cup \{(\bar{\bar{x}}_i,x_j)\} \quad .$$

Instances will be given at the end including plots of the corresponding dual functions.

3. ERROR ESTIMATES

In this section we will develop error estimates for the serendipity projector B. For this purpose we consider the *remainder projectors* $R_s = I-P_s$, $R_t = I-P_t$, $\bar{R}_s = I-\bar{P}_s$, $\bar{R}_t = I-\bar{P}_t$, $\bar{\bar{R}}_s = I-\bar{\bar{P}}_s$ and $\bar{\bar{R}}_t = I-\bar{\bar{P}}_t$. It follows from a general result of Gordon[3] that the remainder projectors generate a *dual lattice* $\mathcal{L}' = \mathcal{L}'(R_s,R_t,\bar{R}_s,\bar{R}_t,\bar{\bar{R}}_s,\bar{\bar{R}}_t)$. Moreover Gordon[3] proved

THEOREM 3 (Duality isomorphism) The mapping $P \leftrightarrow I-P$ is a lattice isomorphism such that

$$PQ \leftrightarrow (I-P)\oplus(I-Q) \quad ,$$

$$P \oplus Q \leftrightarrow (I-P)(I-Q) \quad .$$

In particular the remainder projector of the product projector is given by:

$$I - P_s P_t = R_s \oplus R_t = R_s + R_t - R_s R_t \quad .$$

It follows now from the duality theorem that

$$I - B = I - P_s \bar{\bar{P}}_t \oplus \bar{P}_s \bar{P}_t \oplus \bar{\bar{P}}_s P_t$$

$$= (I - P_s \bar{\bar{P}}_t)(I - \bar{P}_s \bar{P}_t)(I - \bar{\bar{P}}_s P_t)$$

$$= (R_s \oplus \bar{\bar{R}}_t)(\bar{R}_s \oplus \bar{R}_t)(\bar{\bar{R}}_s \oplus R_t) \quad .$$

Taking into account the containment relations we obtain

THEOREM 4 (Serendipity remainder) The remainder of B is given by

$$I - B = \bar{\bar{R}}_s + \bar{R}_s R_t + R_s \bar{R}_t + \bar{\bar{R}}_t - \bar{\bar{R}}_s R_t - \bar{R}_s \bar{R}_t - R_s \bar{\bar{R}}_t \quad .$$

Thus we can apply the error estimates for univariate Lagrange interpolation. For this reason let $C^{(q,r)} := C^{(q,r)}(S)$ be the space of functions f for which all partial derivatives $f^{(k,\ell)}$ $(k \le q\ ,\ \ell \le r)$ are continuous. Furthermore, let

$$\|f\| = \max\{|f(s,t)| \ : \ (s,t) \in S\}$$

and

$$c_{k,\ell} = \|f^{(k+1,\ell+1)}\| / ((k+1)!(\ell+1)!) \quad .$$

The representation of the serendipity remainder of Theorem 4 yields

THEOREM 5 (Serendipity error estimate) Suppose that

$f \in C^{(\bar{\bar{m}}+1,\bar{\bar{m}}+1)}(S)$, then

$$\|f-B(f)\| \le c_{\bar{\bar{m}},0}h^{\bar{\bar{m}}+1} + c_{\bar{m},m}h^{\bar{m}+m+2} + c_{m,\bar{m}}h^{m+\bar{m}+2} + c_{0,\bar{\bar{m}}}h^{\bar{\bar{m}}+1}$$
$$+ c_{\bar{\bar{m}},m}h^{\bar{\bar{m}}+m+2} + c_{\bar{m},\bar{m}}h^{\bar{m}+\bar{m}+2} + c_{m,\bar{\bar{m}}}h^{m+\bar{\bar{m}}+2} .$$

Now let Π_L be the space of bivariate polynomials of total degree less or equal L . Using Theorem 5 we obtain:

THEOREM 6 (Degree of serendipity interpolation) Suppose that $K := \min(m+\bar{m}+2,\bar{\bar{m}}+1)$, then $\Pi_K \subset \mathcal{R}(B)$ and $\Pi_{K+1} \not\subset \mathcal{R}(B)$, i.e. serendipity interpolation produces elements of degree K .

We conclude with a list of *minimal serendipity elements*, i.e. the number of knots, $|\mathcal{S}(B)| = 2(m+1)(\bar{\bar{m}}-m)+(\bar{m}+1)^2$, is minimal for each fixed degree K . Furthermore it can be shown that for each K there exists a unique minimal serendipity element of degree K .

K	m	$\bar{m}$	$\bar{\bar{m}}$	Finite element stencil	$\lvert\mathcal{S}(B)\rvert$	References
1	1	1	1	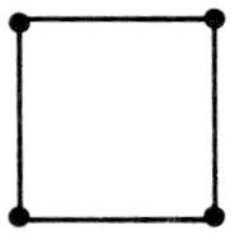	4	2,4,9
2	1	1	2	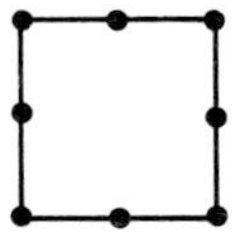	8	1,2,4,9,12
3	1	1	3	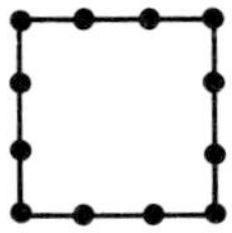	12	1,4,9,12

K	m	$\bar{m}$	$\bar{\bar{m}}$	Finite element stencil	$\|S(B)\|$	References
4	1	2	4	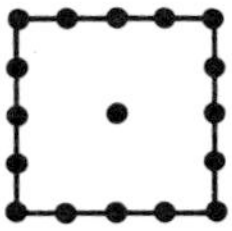	17	2,9,12
5	1	3	5	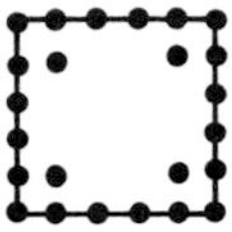	24	

The plots of some dual functions for the minimal elements of degree 4 and 5 are given in Figs. 1 and 2. The other dual functions can be constructed by symmetry arguments. For the plots of minimal elements of degree 2 and 3 we refer to Zienkiewicz[12].

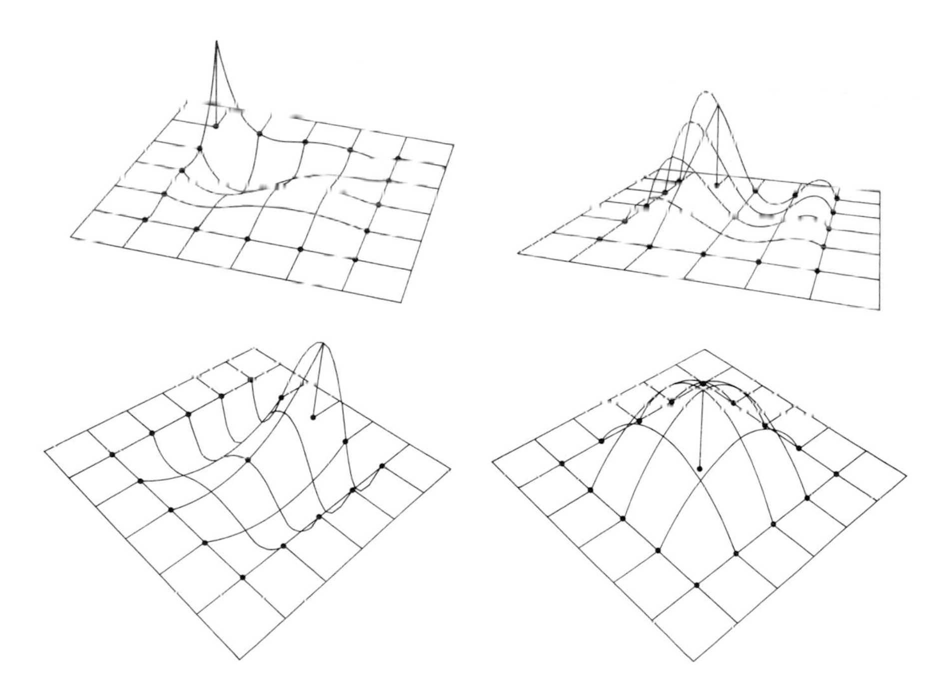

Figure 1 Dual functions for minimal elements of degree 4

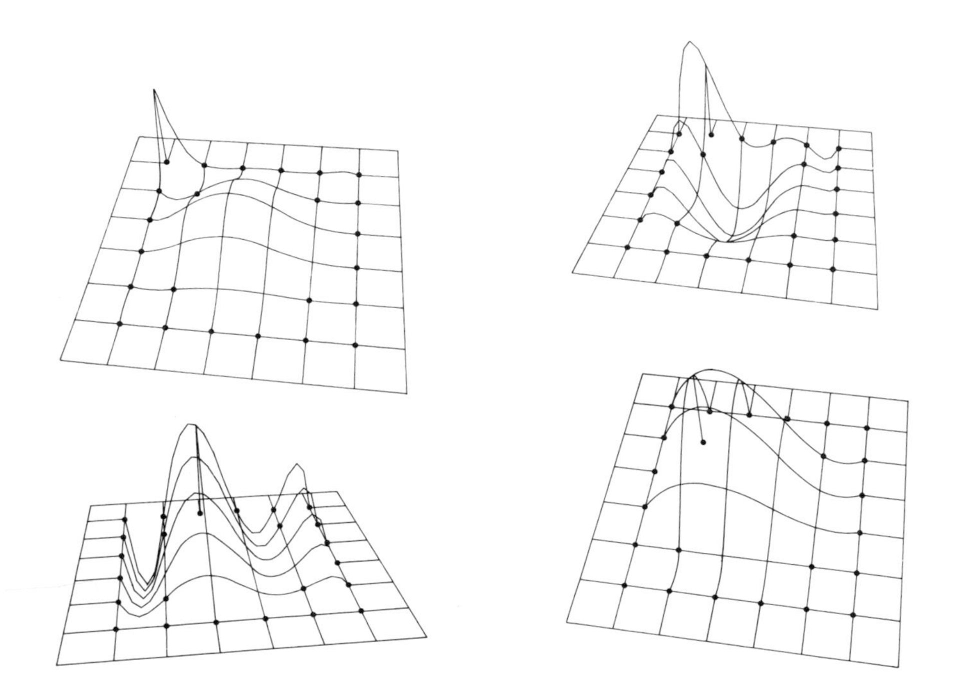

Figure 2 Dual functions for minimal elements of degree 5

4. CONCLUDING REMARKS

The algebraic aspect of our treatment do apply equally to other types of interpolation (cf. [2]). Moreover our approach can be extended to Boolean sums of more than three product projectors. The algebraic details are given in [2]. The connection between rectangular locally minimal elements of Lancaster and Watkins[9] and n-th order blending will be investigated in a forthcoming paper.

ACKNOWLEDGEMENTS

We wish to express our thanks to Prof. Dr. E. Ehlich,

Bochum, who enables the realization of our examples at TR 440 - Rechenzentrum of the University of Bochum and to cand.math. G. Baszenski for generating the plot programs.

REFERENCES

1. Barnhill, R.E. (1976). Blending Function Interpolation: A Survey and Some New Results, *in* "Numerische Methoden der Approximationstheorie Bd. 3" (Eds. L. Collatz, H. Werner, G. Meinardus), ISNM Vol. 30, 43-89.

2. Delvos, F.-J. and Posdorf, H. (1977). N-th Order Blending, *in* "Constructive Theory of Functions of Several Variables" (Eds. W. Schempp, K. Zeller) Lecture Notes in Mathematics 571, 53-64.

3. Gordon, W.J. (1969). "Distributive Lattices and Approximation of Multivariate Functions", Proc. Symp. Approximation with Special Emphasis on Spline Functions (Madison, Wisc., 1969) (Ed. I.J. Schoenberg) 223-277, Academic Press, New York.

4. Gordon, W.J. (1971). Blending-function Methods of Bivariate and Multivariate Interpolation and Approximation, *S.I.A.M. Jour. Num. An.* 8, No.1, 158-177.

5. Gordon, W.J. and Hall, C.A. (1973). Transinite Element Methods: Blending Function Methods over Arbitrary Curved Element Domains, *Num. Math.* 21, 109-129.

6. Gordon, W.J. and Wixom, J.A. (1974). Pseudo-harmonic Interpolation on Convex Domains, *S.I.A.M. Jour. Num. An.* 11, 909-933.

7. Hall, C.A. (1976). Transfinite Interpolation and Application to Engineering Problems, *in* "Theory of Approximation with Applications" (Eds. A.G. Law, B.N. Sahney) 308-331, Academic Press, New York.

8. Lancaster, P. (1976). Interpolation in a Rectangle and Finite Elements of High Degree, *J. Inst. Maths. Applics.* 18, 67-77.

9. Lancaster, P. and Watkins, D.S. (1976). Interpolation in the Plane and Rectangular Finite Elements, *in* "Numerische Behandlung von Differentialgleichungen Bd. 2" (Eds. J. Albrecht, L. Collatz) ISNM Vol. 31, 125-145.

10. Watkins, D.S. (1976). A Conforming Rectangular Plate Element, *in* "The Mathematics of Finite Elements and

Applications II, MAFELAP 1975" (Ed. J.R. Whiteman) 77-83, Academic Press.

11. Watkins, D.S. (1976). On the Construction of Conforming Rectangular Plate Elements, *Int. J. Num. Meth. Eng.* 10, 925-933.

12. Zienkiewicz, O.C. (1971). The Finite Element Method in Engineering Science, McGraw-Hill, London.

THE ALGEBRAIC STRUCTURE OF LINEAR SYSTEMS OCCURRING IN NUMERICAL METHODS FOR SPLINE INTERPOLATION

G. Meinardus

University of Siegen
Siegen, Germany

1.

The classical splines are understood as piecewise polynomials belonging to a partition of the given set and to some differentiability class. The main question for almost all kinds of problems in this field seems to be the one of constructing a suitable basis for those spline spaces. In one real variable this problem seems to be solved completely by considering the so-called B-splines[7,8]. These functions include in their constructive definition all the continuity conditions of the spline and of its derivatives. It is well known, however, that in the case of splines of a low order the more elementary approach is simpler and easier to handle instead of using those B-splines. There one tries to formulate the differentiability conditions as linear relations for the spline and some of its derivatives at the knots. Usually[1,5] the values of these functions will occur in those relations at $m-1$ successive knots, m being the order of the spline. Here we will consider the possibility of using only three successive points. The price one has to pay for this advantage is the increased order of the linear systems. We will have to consider, however, matrices of a block structure.

To succeed in this way one needs to construct linear functionals of a special type, annihilating some polynomial spaces. We will show in two essential instances how this can be done (for the cubic case cp.[4]). Furthermore we will give some remarks on special cases in two variables which may be treated analogously.

2.

Let K be a set of knots x_υ :

$$a = x_0 < x_1 < \dots < x_n = b$$

and let m be a natural number not less than 2 . We call a real function s on $[a,b]$ a spline function of order m belonging to the set K if

1. $s \in C^{m-2}[a,b]$

and

2. the restriction of s to the interval

$$I_\upsilon = [x_{\upsilon-1}, x_\upsilon] \ , \quad \upsilon = 1,2,\dots,n \ ,$$

belongs to the space π_{m-1} of polynomials of degree at most $m-1$:

$$s|I_\upsilon = p_\upsilon \in \pi_{m-1} \ .$$

The first property is obviously equivalent to some connection for the polynomials p_υ : there exist numbers γ_υ , $\upsilon = 1,2,\dots,n-1$, such that

$$p_{\upsilon+1}(x) = p_\upsilon(x) + \gamma_\upsilon(x-x_\upsilon)^{m-1}$$

holds. This again is equivalent to the statement

$$p_{\upsilon+1}^{(\mu)}(x_\upsilon) = p_\upsilon^{(\mu)}(x_\upsilon) \tag{2.1}$$

for $\upsilon = 1,2,\ldots,n-1$; $\mu = 0,1,\ldots,m-2$.

From (2.1) we may proceed in the following way: We will express higher derivatives of p_υ and $p_{\upsilon+1}$ at $x = x_\upsilon$ by lower derivatives and function values. Because of the first property of s one has to compare these expressions. This leads to the possibility to construct afterwards the polynomials p_υ by a simple HERMITE interpolation method. - It is useful to handle the cases of an even and of an odd order separately.

3.

Let $m = 2k+2$, $k \geq 1$ (the case $k = 0$ is here without interest).

LEMMA There exist unique numbers α_r^μ and β_r^μ , $r = 0,1,\ldots,k$; $\mu = 1,2,\ldots,k$, such that

$$p^{(k+\mu)}(0) = \sum_{r=0}^{k} \alpha_r^\mu p^{(r)}(0) + \sum_{r=0}^{k} \beta_r^\mu p^{(r)}(1) \tag{3.1}$$

is valid for every $p \in \pi_{2k+1}$. Consider the polynomials

$$v_\mu(\tau) = \sum_{r=0}^{k} \alpha_r^\mu \tau^r , \quad w_\mu(\tau) = \sum_{r=0}^{k} \beta_r^\mu \tau^r .$$

Then

$$v_\mu(\tau) + e^\tau w_\mu(\tau) = \tau^{k+\mu} + \sigma(\tau^{2k+2}) \tag{3.2}$$

for $\tau \to 0$.

Proof The existence and uniqueness of the $\alpha_r^\mu, \beta_r^\mu$ with the property (3.1) follows from the existence and uniqueness of an obvious HERMITE interpolation problem. From this remark one could derive explicit expressions for those numbers (cp. [6]). - To prove (3.2) we consider the linear functional

$$L(y) = y^{(k+\mu)}(0) - \sum_{r=0}^{k} \alpha_r^\mu y^{(r)}(0) - \sum_{r=0}^{k} \beta_r^\mu y^{(r)}(1)$$

for the function

$$y = y(t) = e^{\tau t} \quad ,$$

τ being a parameter. The Taylor expansion of $L(y)$ with respect to τ at $\tau = 0$ starts with the $(2k+2)^{nd}$ power of τ, which is equivalent to (3.2).

To use the representation (3.1) not only for the points 0 and 1 but for any two different numbers ξ, η we transform the interval $0 \le t \le 1$ to the interval with the endpoints ξ, η by

$$x = \xi + t(\eta - \xi)$$

and define

$$q(x) = p(t) \quad \text{and} \quad h = \eta - \xi \quad .$$

Then the formula

$$q^{(k+\mu)}(\xi) = \sum_{r=0}^{k} \alpha_r^\mu h^{r-\mu-k} q^{(r)}(\xi) + \sum_{r=0}^{k} \beta_r^\mu h^{r-\mu-k} q^{(r)}(\eta) \qquad (3.3)$$

holds for every $q \in \pi_{2k+1}$. With the abbreviation

$$\lambda(x) = \begin{pmatrix} q'(x) \\ q''(x) \\ \vdots \\ q^{(k)}(x) \end{pmatrix}$$

the formula (3.3) yields

$$\begin{pmatrix} q^{(k+1)}(\xi) \\ q^{(k+2)}(\xi) \\ \vdots \\ q^{(2k)}(\xi) \end{pmatrix} = q(\xi)\begin{pmatrix} \alpha_0^1 h^{-1-k} \\ \alpha_0^2 h^{-2-k} \\ \vdots \\ \alpha_0^k h^{-2k} \end{pmatrix} + q(\eta)\begin{pmatrix} \beta_0^1 h^{-1-k} \\ \beta_0^2 h^{-2-k} \\ \vdots \\ \beta_0^k h^{-2k} \end{pmatrix} +$$

$$+ h^{k} H^{-1} AH\lambda(\xi) + h^{-k} H^{-1} BH\lambda(\eta) \quad , \qquad (3.4)$$

where the square matrices H, A and B are of type (k,k) defined by

$$H = H(h) = \mathrm{diag}(h, h^2, \ldots, h^k) \quad ,$$

$$A = ((\alpha_r^\mu)) \ , \quad B = ((\beta_r^\mu)) \ , \quad r,\mu = 1,2,\ldots,k \quad .$$

Now we choose, for $\nu = 1,2,\ldots,n-1$,

1. $\xi = x_\nu$, $\eta = x_{\nu-1}$, $h = -h_\nu$, $q = p_\nu$

and

2. $\xi = x_\nu$, $\eta = x_{\nu+1}$, $h = h_{\nu+1}$, $q = p_{\nu+1}$.

Here p_ν and $p_{\nu+1}$ are the restrictions of s to the intervals I_ν , $I_{\nu+1}$ respectively. We have

$$s^{(k+\mu)}(x_\nu) = p_\nu^{(k+\mu)}(x_\nu) = p_{\nu+1}^{(k+\mu)}(x_\nu)$$

for $\nu = 1,2,\ldots,n-1$ and $\mu = 1,2,\ldots,k$. With the obvious property:

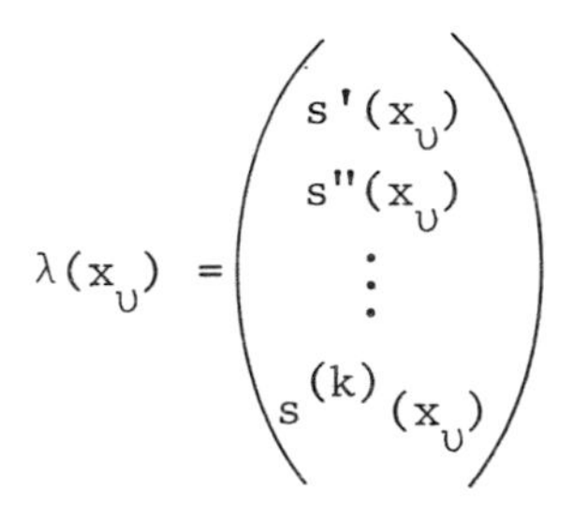

$$\lambda(x_\nu) = \begin{pmatrix} s'(x_\nu) \\ s''(x_\nu) \\ \vdots \\ s^{(k)}(x_\nu) \end{pmatrix}$$

we eventually are led to the relation

$$\begin{aligned} A_\nu \lambda(x_\nu) + B_{1\nu}\lambda(x_{\nu-1}) + B_{2\nu}\lambda(x_{\nu+1}) &= \\ &= a_\nu s(x_\nu) + b_{1\nu}s(x_{\nu-1}) + b_{2\nu}s(x_{\nu+1}) \end{aligned} \tag{3.5}$$

for $\nu = 1,2,\ldots,n-1$. Here we have used the abbreviations

$$A_\nu = (-h_\nu)^{-k}H^{-1}(-h_\nu)AH(-h_\nu) - (h_{\nu+1})^{-k}H^{-1}(h_{\nu+1})AH(h_{\nu+1}) \; ,$$

$$B_{1\nu} = (-h_\nu)^{-k}H^{-1}(-h_\nu)BH(-h_\nu) \; ,$$

$$B_{2\nu} = -(h_{\nu+1})^{-k}H^{-1}(h_{\nu+1})BH(h_{\nu+1}) \; ,$$

$$a_\nu = \begin{pmatrix} \alpha_0^1(h_{\nu+1}^{-k-1} - (-h_\nu)^{-k-1}) \\ \vdots \\ \alpha_0^k(h_{\nu+1}^{-2k} - (-h_\nu)^{-2k}) \end{pmatrix} ,$$

$$b_{1\nu} = -\begin{pmatrix} \beta_0^1(-h_\nu)^{-k-1} \\ \vdots \\ \beta_0^k(-h_\nu)^{-2k} \end{pmatrix} , \quad b_{2\nu} = \begin{pmatrix} \beta_0^1(h_{\nu+1})^{-k-1} \\ \vdots \\ \beta_0^k(h_{\nu+1})^{-2k} \end{pmatrix} .$$

The linear relation (3.5) is equivalent to the property that the spline s belongs to C^{2k} at a neighbourhood of the point x_ν. It links the values of s and its derivatives up to the order k at three successive knots. If we have in mind the interpolation of some function at the knots and if we are given boundary conditions, e.g. natural or periodic, we have the well known block structure in the full matrix. To look for the condition of these matrices is important for numerical purposes. It is our intention to deal with this question in another paper.

The cubic case, $k=1$, is very familiar. There we have just one relation for every ν, $\nu=1,2,\ldots,n-1$. The matrix of the corresponding interpolation problem, using periodic boundary conditions, is equal to

$$M(2I+(I-P)T+PT^{-1}) \tag{3.6}$$

with the matrices M, P, T of type (n,n) given by

$$M = 2\,\mathrm{diag}\left(\frac{1}{h_\nu}+\frac{1}{h_{\nu+1}}\right), \quad P = \mathrm{diag}\,\delta_\nu$$

where

$$\delta_\nu = \frac{h_\nu}{h_\nu+h_{\nu+1}}, \quad \nu=1,2,\ldots,n,$$

and

$$h_{n+1} = h_1 .$$

Furthermore we have the permutation matrix

$$T = ((t_{\nu\mu})), \quad \nu,\mu=1,2,\ldots,n$$

with

$$t = \begin{cases} 1 & \text{if } \upsilon-\mu \equiv 1 \bmod n \ , \\ 0 & \text{otherwise} \ . \end{cases}$$

The positive numbers δ_υ satisfy the restriction

$$\delta_1\delta_2\ldots\delta_n = (1-\delta_1)(1-\delta_2)\ldots(1-\delta_n) \quad .$$

It turns out that there is a factorization of the matrix in (3.6): There exists a diagonal matrix

$$D = \operatorname{diag} d_\upsilon \ , \quad d_\upsilon > 0 \ ,$$

such that with

$$X = D^{-1}(I-P)$$

and

$$Y = D^{-1}P$$

the identity

$$M(2I+(I-P)T+PT^{-1}) = MD(I+XT)(I+YT^{-1})$$

holds. The numbers d_ν are uniquely determined by the formula

$$d_\upsilon = 2 - \frac{\delta_{\upsilon-1}(1-\delta_\upsilon)}{d_{\upsilon-1}} \quad . \tag{3.7}$$

This can be interpreted as a periodic continued fraction if we define for all integer υ,μ :

$$\delta_\upsilon = \delta_\mu \quad \text{for} \quad \upsilon \equiv \mu \bmod n$$

and look for a solution of (3.7) with

$$d_\upsilon = d_\mu \quad \text{for} \quad \upsilon \equiv \mu \bmod n \quad .$$

It would be of importance, not only in the case of periodic boundary conditions, to have an analogue of this factorization of the corresponding matrices.

4.

Let $m = 2k+1$, $k \geq 1$. We will only sketch the analysis in this even degree case. Having in mind the interpolation at the mid-points

$$\frac{x_{\upsilon-1} + x_\upsilon}{2} \quad , \qquad \upsilon = 1,2,\dots,n$$

of the intervals I_υ it is natural to ask for a representation

$$p^{(k+\mu)}(0) = \gamma^\mu p(\tfrac{1}{2}) + \sum_{r=0}^{k-1} \tilde{\alpha}^\mu_r p^{(r)}(0) + \sum_{r=0}^{k-1} \tilde{\beta}^\mu_r p^{(r)}(1) \quad , \qquad (4.1)$$

for $\mu = 0,1,\dots,k-1$, which is valid for every $p \in \pi_{2k}$. Such numbers $\gamma^\mu, \tilde{\alpha}^\mu_r, \tilde{\beta}^\mu_r$ exist uniquely. Considering the polynomials

$$\tilde{\upsilon}_\mu(\tau) = \sum_{r=0}^{k-1} \tilde{\alpha}^\mu_r \tau^r \quad , \quad \tilde{w}_\mu(\tau) = \sum_{r=0}^{k-1} \tilde{\beta}^\mu_r \tau^r$$

the following asymptotic relation holds:

$$\gamma^\mu e^{\tau/2} + \tilde{\upsilon}_\mu(\tau) + e^\tau \tilde{w}_\mu(\tau) = \tau^{k+\mu} + O(\tau^{2k+1}) \qquad (4.2)$$

for $\tau \to 0$.

Now everything is quite analogous to the procedure in section 3.

5.

In multivariate cases one always has the fact that continuity conditions for the spline function and some of its derivatives may lead to algebraic difficulties and are very restrictive, too, unless one uses the tensor product approach. One could instead think of giving up the continuity conditions but to consider some linear relations which are due to the demand that some derivatives coincide at points which determine the partition. We sketch the idea in a simple case: In the x-y plane let us consider a triangulation of the given set by equilateral triangles. In every triangle the function s ("spline") may be represented by a member of the linear space

$$Z = \operatorname{span}(1, \operatorname{Re} z, \operatorname{Im} z, \ldots, \operatorname{Re} z^4, \operatorname{Im} z^4)$$

of dimension 9, $z = x+iy$. On the edges of these triangles one may define s in some special way. Of course, not even continuity of s in the whole set is provided. - The interpolation problem for a given function g could consist in giving just the values of g at the corners of the partition in addition to some linear relations for s. One may, as an example, try to express for $u \in Z$ the derivative u_{xx} at some of the corners by a linear combination of u, u_x, u_y at the three corners. Transforming the triangle to the one with the corners

$$(1,0)\ ,\ (-\tfrac{1}{2}, \tfrac{1}{2}\sqrt{3})\ ,\ (-\tfrac{1}{2}, -\tfrac{1}{2}\sqrt{3})$$

this means to find the numbers $\alpha_1, \alpha_2, \alpha_3, \beta_1, \beta_2, \beta_3$ and

$\gamma_1, \gamma_2, \gamma_3$ such that

$$u_{xx}(1,0) = \alpha_1 u(1,0) + \alpha_2 u(-\frac{1}{2}, \frac{1}{2}\sqrt{3}) + \alpha_3 u(-\frac{1}{2}, -\frac{1}{2}\sqrt{3}) +$$

$$+ \beta_1 u_x(1,0) + \beta_2 u_x(-\frac{1}{2}, \frac{1}{2}\sqrt{3}) + \beta_3 u_x(-\frac{1}{2}, -\frac{1}{2}\sqrt{3}) +$$

$$+ \gamma_1 u_y(1,0) + \gamma_2 u_y(-\frac{1}{2}, \frac{1}{2}\sqrt{3}) + \gamma_3 u_y(-\frac{1}{2}, -\frac{1}{2}\sqrt{3})$$

is valid for every $u \in Z$. In fact, it is easy to compute these coefficients. Even the techniques of the foregoing sections, using the exponential function, may be applied. There are possibilities to use spherical harmonics in higher dimensions, too. So, for three variables, it is likely that in partitions consisting of equilateral tetrahedra the spherical harmonics up to the degree 3 are useful.

REFERENCES

1. Ahlberg, J.H., Nilson, E.N. and Walsh, J.L. (1967). "The Theory of Splines and Their Applications", Academic Press, New York and London.
2. Curry, H.B. and Schoenberg, I.J. (1966). On Polya Frequency Functions IV, *J. Anal. Math.* 17, 71-107.
3. De Boor, C. (1972). On Calculating with B-Splines, *J. Approximation Theory*, 6, 50-62.
4. Merz, G. (1974). Splines, *Überblicke Mathematik* 7, 115-165.
5. Schurer, F. (1970). "A Note on Interpolating Periodic Quintic Spline Functions", Proc. Symp. Lancaster, 71-81 (Ed. A. Talbot).
6. Stancu, D.D. (1957). A supra Formulei de Interpolare a lui Hermite si a unor Aplicatii ale Acestei. *Studii si Cercetari de Mat. (Cluj)* VIII, 339-355.

INTEGRAL REPRESENTATION OF REMAINDER FUNCTIONALS IN ONE AND SEVERAL VARIABLES

Harold S. Shapiro

Institution för Matematik
Kungl. Tekniska Högskolan
Stockholm, Sweden

1. INTRODUCTION

Let us illustrate the kind of problem here under consideration by means of a rather trivial example. Suppose we approximate the integral $\int_0^1 f(x)dx$ of a function, using the simplest case of the "trapezoidal rule", by $\frac{1}{2}(f(0) + f(1))$. To study the error we introduce the *remainder functional*

$$Rf = \int_0^1 f(x)dx - \tfrac{1}{2}(f(0) + f(1)) \quad .$$

Since R vanishes on polynomials of degree not exceeding 1 , one guesses that Rf is representable, for smooth f , in the form of an integral

$$Rf = \int_0^1 f''(x)k(x)dx \tag{1.1}$$

where k is some integrable function k . That this is indeed so (it is a special case of a theorem of Peano) can be seen as follows: write (1.1) as a differential equation in the sense of distributions:

$$D^2k = R \equiv \chi_{[0,1]} - \tfrac{1}{2}\delta_{(0)} - \tfrac{1}{2}\delta_{(1)}$$

where $\chi_{[0,1]}$ is the characteristic function of $[0,1]$, $\delta_{(0)}$, $\delta_{(1)}$ denote Dirac functionals, and D is the distributional derivative. Integrating once gives

$$Dk = \begin{cases} x - \frac{1}{2} & \text{on } [0,1] \\ 0 & \text{outside } [0,1] \end{cases}$$

and another integration gives

$$k(x) = \begin{cases} \int_0^x (t - \frac{1}{2})dt = -\frac{1}{2}x(1-x) & \text{on } [0,1] \\ 0 \quad \text{outside } [0,1] \end{cases}$$

It is important to observe that two integrations of R produce a function k *supported in* $[0,1]$, and this is a consequence of R annihilating linear polynomials; attempts to represent Rf as $\int f'''(x)h(x)dx$ would require another integration and give a function $h(x)$ not of compact support; while the resulting formula would be valid for suitably restricted f, it would have little interest from the standpoint of approximation theory.

In several variables things are more complex: suppose for instance Ω is a nice domain in $\mathbb{R}^2$, and we approximate $\iint_\Omega f dxdy$ by simply taking $Af(x_o,y_o)$ where A is the area of Ω and (x_o,y_o) its centroid. The associated remainder functional is now

$$Rf = \iint_\Omega f dxdy - Af(x_o,y_o)$$

and vanishes on all polynomials of degree at most one. Again, this suggests the existence of a representation formula

$$Rf = \iint_\Omega (f_{xx}k_1 + f_{xy}k_2 + f_{yy}k_3)dxdy$$

for suitably chosen integrable functions k_1,k_2,k_3 (depending only on R), valid for all smooth functions f . So far as I am aware there is no theorem in the literature which tells us this, however. It is a corollary of Theorem 3 below that such k_i can be chosen, and indeed in the class $L^p(\Omega)$ for any given $p > \infty$. The k_i here are not unique. What is worse, there is no known way of finding them, the proof being non-constructive. Most of what is currently known about "cubature formulas" may be found in the books[7,8,10].

2. THE ONE-VARIABLE CASE

Although the one-variable case is completely clarified by Peano's theorem (cf. Davis[2], p.69) we review it again here, from a *functional-analytic* standpoint, inasmuch as this standpoint appears to us *essential* for the later work in several variables.

Let Ω denote an open bounded interval of the real line $\mathbb{R}$, and $\bar{\Omega}$ its closure. By $C^n(\bar{\Omega})$ we denote the class of functions f on Ω whose derivatives of order up to and including n exist and are uniformly continuous on Ω , hence continuously extendible to $\bar{\Omega}$ (we always assume them so extended). This is a Banach space if we adopt the norm

$$\|f\|_{C^n(\bar{\Omega})} = \sum_{j=0}^{n} \|f^{(j)}\|_{C(\bar{\Omega})}$$

where of course $\|\cdot\|_{C(\bar{\Omega})}$ denotes the max norm on $\bar{\Omega}$. By means of the map J defined thus:

$$J: f \to (f, f', \ldots, f^{(n)})$$

$C^n(\bar{\Omega})$ is embedded isometrically into $C(\bar{\Omega})^{n+1}$, therefore every continuous linear functional on $C^n(\bar{\Omega})$ may be identified with the restriction to the range of J of a continuous linear functional on $C(\bar{\Omega})^{n+1}$, that is with an (n+1)-tuple of bounded measures on $\bar{\Omega}$. Thus, to every R in $C^n(\bar{\Omega})^*$, the space of bounded linear functionals on $C^n(\bar{\Omega})$, there exists a (non-unique) set $(\mu_0, \mu_1, \ldots, \mu_n)$ of elements of $M(\bar{\Omega})$ (bounded measures with support in $\bar{\Omega}$) such that

$$Rf = \sum_{j=0}^{n} \int f^{(j)} d\mu_j \quad , \qquad \forall f \in C^n(\bar{\Omega}) \quad . \tag{2.1}$$

If we adopt the standpoint of distribution theory, we may rewrite this as

$$R = \sum_{j=0}^{n} (-D)^j \mu_j \tag{2.2}$$

where D denotes the distributional derivative, and the measures μ_j and the functional R are interpreted as distributions in $\mathbb{R}$ with support in $\bar{\Omega}$. Thus, every element of $C^n(\bar{\Omega})^*$ is (non-uniquely, if $n \geq 1$) a sum of elements of $M(\bar{\Omega})$ and derivatives thereof of orders not exceeding n, and conversely, every such sum is an element of $C^n(\bar{\Omega})^*$. All this is of course standard functional analysis, which we recall here mainly for notational reasons.

The property of a functional $R \in C^n(\bar{\Omega})^*$ that is typical of a "remainder" is that it annihilates polynomials of low degree. This circumstance may be exploited to simplify the representation (2.2), and that is the essence of *Peano's theorem*. Let $\mathcal{P}_k$ denote the set of polynomials of degree at most k, then one way to formulate it is:

THEOREM 1 (PEANO) *Let* $R \in C^n(\bar{\Omega})^*$ *satisfy* $R\mathcal{P}_{n-1} = 0$. *Then there exists a unique measure* $\mu \in M(\Omega)$ *such that* $R = (-D)^n \mu$ *or, what is the same*

$$R_f = \int f^{(n)} d\mu \quad , \quad \forall f \in C^n(\bar{\Omega}) \quad . \tag{2.3}$$

Remark 1 To be strictly consistent with the distributional interpretation, we should say (2.3) holds only for $f \in C^\infty(\mathbb{R})$ with compact support, however this class (restricted to $\bar{\Omega}$) is dense in $C^n(\bar{\Omega})$, so we can in (2.3) allow f in the latter class, more natural in approximation theory.

Remark 2 In the classical version of Peano's theorem, R is assumed to be a distribution of order at most $n-1$ (that is, an element of $C^{n-1}(\bar{\Omega})^*$) rather than n , as in the above formulation, and that enables one to get, in place of (2.3), the nicer representation $Rf = \int f^{(n)}(x)k(x)dx$, where k is a function of bounded variation vanishing outside $\bar{\Omega}$. More will be said about this point below.

The usual proof of Peano's theorem is to develop f in a Taylor series up to terms of degree $n-1$, with an integral remainder term, and then apply R to both sides. This method involves a difficulty about "taking R inside the integral sign" which can be treated rigorously, although it is glossed over in some accounts. We sketch here two alternative proofs of the theorem. Observe first that the *uniqueness* part is trivial, since if μ_1, μ_2 are two solutions of $(-D^n)\mu = R$ with support in $\bar{\Omega}$, then $D^n(\mu_1 - \mu_2) = 0$ so that $\mu_1 - \mu_2$ is a polynomial with compact support, hence zero. Hence we consider henceforth only the existence of μ .

Proof # 1 of Theorem 1 Without loss of generality, let $\Omega = (-\tau, \tau)$. We claim first that the distributional differential equation $D^n u = R$ has a (unique) distributional solution u with support in $\bar{\Omega}$. Indeed, taking Fourier transforms, we get $(-ix)^n \hat{u}(x) = \hat{R}(x)$. Now, $\hat{R}$ is an entire function of exponential type at most τ, of polynomial growth (indeed, $O(|x|^n)$) on the real axis. The hypothesis $RP_{n-1} = 0$ implies that $\hat{R}(x)$ has a zero of order n (at least) at the origin, hence $\hat{R}(x)/(-ix)^n$ is an entire function of exponential type at most τ, bounded on the real axis. By one form of the Paley-Wiener theorem, this function is the Fourier transform of a tempered distribution u with support in $\bar{\Omega}$, and $D^n u = R$. To complete the proof it suffices to show that u is a locally finite measure on $\mathbb{R}$, since we already know that its support is in $\bar{\Omega}$. From (2.2) we have

$$D^n u = \sum_{1=0}^{n} (-D)^j \mu_j \quad , \quad \mu_j \in M(\bar{\Omega}) \quad . \tag{2.4}$$

Now, each term on the right is certainly the nth derivative of a locally finite measure on $\mathbb{R}$ (not necessarily bounded, nor compactly supported), so we have $D^n u = D^n \nu$, where ν is a locally finite measure on $\mathbb{R}$. Thus $u = \nu + p$, where $p \in P_{n-1}$ and so u is a locally finite measure on $\mathbb{R}$, completing the proof.

Remark Observe that, in the last stage of the proof, if we had assumed more about R, e.g. that it were in $C^{n-1}(\bar{\Omega})^*$, then the summation on the right of (2.4) would only go up to $n-1$ and thus ν would be a function of bounded variation on $\mathbb{R}$, whence u would be a function of bounded variation supported in $\bar{\Omega}$. This is the classical form of Peano's theorem. (Similarly, $R \in C^{n-2}(\Omega)^*$ gives u as the integral of a function of bounded variation, etc.)

Proof # 2 of Theorem 1 Consider the map

$$T : f \rightarrow f^{(n)}$$

from $C^n(\bar{\Omega})$ to $C(\bar{\Omega})$. Its kernel is P_{n-1} and its range is $C(\bar{\Omega})$. Its adjoint T^* goes from $C(\bar{\Omega})^* = M(\bar{\Omega})$ to $C^n(\bar{\Omega})^*$ in accordance with the scheme

$$\langle Tf,\mu\rangle = \langle f,T^* \mu\rangle$$

where here $\langle , \rangle$ denotes the generic pairing between a linear space and its dual, and μ an arbitrary element of $M(\bar{\Omega})$. That is, $T^*\mu$ is that functional on $C^n(\bar{\Omega})$ which maps f to $\int f^{(n)} d\mu$. Therefore, it is sufficient to prove that the given functional R belongs to the range of T^*. Now, T has a closed range (indeed, it is surjective!), hence by a general theorem of functional analysis (see, for example [3], p.487) the range of T^* is precisely the annihilator of the kernel of T. Since, by hypothesis, R annihilates this kernel, it is in the range of T^*, which completes the proof.

Remark 1 As we shall see, this proof extends with only notational changes, to several variables. Unfortunately, it is a pure existence proof. (The first proof is constructive in that it gives explicitly the Fourier transform of k, but this proof seems not extendible to several variables.)

Remark 2 Radon[5] proved a generalization of Peano's theorem wherein one replaces $f^{(n)}$ by Lf, with L an nth order linear differential operator (with variable coefficients), and P_{n-1} by the vector space of solutions of $Lu = 0$. This version is also a consequence of the method of proof # 2 above.

Remark 3 Let us look again at the representation

$$Rf = \int_{\Omega} f^{(n)}(x)k(x)dx \quad , \qquad f \in C^n(\bar{\Omega}) \tag{2.5}$$

with k of bounded variation, valid whenever $R \in C^{n-1}(\bar{\Omega})^*$ and $RP_{n-1} = 0$. Knowing its existence, we can compute k as follows: fix $\xi \in \Omega$, and choose for f a solution of the equation $f^{(n)} = \delta_{(\xi)}$, namely the spline

$$f_{\xi}(x) = \frac{(x-\xi)_+^{n-1}}{(n-1)!} \quad . \tag{2.6}$$

Then $k(\xi)$ is equal to R evaluated at f_{ξ} . At first sight this reasoning seems merely formal, because f_{ξ} is of class C^{n-2} and not C^n as required in (2.5). However, in all cases of practical interest R has, in the sense of distributions, locally the structure of a smooth function, except for some finite set S of points, and the same is true of k which arises from R by integration. Consider now a point ξ in $\Omega \backslash S$. Then both $k(\xi)$ and Rf_{ξ} are well defined, the latter by the usual methods of distribution theory, i.e. split R into a smooth function supported near ξ , and a distribution vanishing on a neighbourhood of ξ , each of which can be evaluated naturally at f_{ξ} . That Rf_{ξ} thus defined is equal to $k(\xi)$ can now be shown rigorously by a straightforward approximation argument, which we omit. This formula for $k(\xi)$ is also due to Peano.

3. SEVERAL VARIABLES

Let Ω denote a bounded open set in $\mathbb{R}^d$. We shall assume for convenience that Ω is *convex* but this is not essential (see the remark below). One defines the Banach space $C^n(\bar{\Omega})$ exactly as in the one-dimensional case, and establishes that $C^n(\bar{\Omega})^*$ consists exactly of those distributions that can be written in the form

$$R = \sum_{|\alpha| \le n} D^{\alpha}\mu_{\alpha} \tag{3.1}$$

where the usual multi-index conventions are employed, and the μ_{α} are bounded measures on $\mathbb{R}^d$ with support in $\bar{\Omega}$. The μ_{α} in this formula are not unique. Sard[8,9] proved the following generalization of Peano's theorem in which all but the highest order terms in (3.1) are eliminated:

THEOREM 2 (SARD) *Let* $R \in C^n(\bar{\Omega})^*$ *satisfy* $RP_{n-1} = 0$. *Then there exist measures* $\mu_{\alpha} \in M(\bar{\Omega})$, $|\alpha| = n$ *such that*

$$R = \sum_{|\alpha|=n} D^{\alpha}\mu_{\alpha} \quad ,$$

in other words such that

$$Rf = (-1)^n \sum_{|\alpha|=n} \int (D^{\alpha}f)d\mu_{\alpha} \quad , \quad \forall f \in C^n(\bar{\Omega}) \quad . \tag{3.2}$$

Remark Sard actually proves a more general representation formula, for R which do not necessarily annihilate P_{n-1}, but that is immediately deducible from the above. We sketch a proof somewhat different from Sard's, modelled on our proof # 2 of Peano's theorem. Consider the map

$$T: f \to (f^{(\alpha_1)}, f^{(\alpha_2)}, \ldots, f^{(\alpha_m)})$$

from $C^n(\bar{\Omega})$ to $C(\bar{\Omega})^m$, where $\alpha_1, \ldots, \alpha_m$ is some ordering of the multi-indices of rank n. Its kernel is P_{n-1}. Its adjoint T^* maps each m-tuple of elements of $M(\bar{\Omega})$ to an element of $C^n(\bar{\Omega})^*$, and an examination of this relationship shows that each R in the range of T^* admits a representation of (3.2). Thus, it is enough to show that the range of T^* includes the annihilator of the kernel of T and, as above, it suffices for this to check that range T is closed. This is equivalent to the existence of a constant A such that

$$\operatorname{dist}(f, \ker T)_{C^n(\bar{\Omega})} \leq A \| Tf \|_{C(\bar{\Omega})^m} ,$$

that is,

$$\operatorname{dist}(f, \mathcal{P}_{n-1})_{C^n(\bar{\Omega})} \leq A \underset{|\alpha|=n}{} \| f^{(\alpha)} \|_\infty . \tag{3.3}$$

We have, then, to verify the estimate (3.3) (which is, in fact, *equivalent* to Sard's theorem). This verification is a simple application of Taylor's formula with remainder, which we leave to the reader.

In Theorem 2, we cannot assert more about the μ_α than that they are measures, because we have not assumed very much about R. It is natural to hope that the underdetermined partial differential equation

$$R = \sum_{|\alpha|=n} D^\alpha u_\alpha$$

has solutions u_α which are in some sense "n integrations smoother" than R itself. Indeed this is certainly true, and easily proven, for *any* distribution R, if we do not impose *restrictions on the supports of the* u_α which, however, is the case here. We shall prove only one typical theorem of the desired kind, in which we suppose R is a *measure*, which corresponds to the important class of remainder functionals that arise in "cubature formulas". For notational reasons we write $d\mu$ in place of R.

THEOREM 3 *Let* $\mu \in M(\bar{\Omega})$ *and* $\int f d\mu = 0$ *for every* $f \in \mathcal{P}_{n-1}$.

Assume $n \le d$. *Then, for every* $p < \frac{d}{d-n}$ *there exists a set of functions* $\{u_\alpha : \text{all } \alpha, \; \alpha = n\}$ *in* $L^p(\Omega)$ *such that*

$$\int f d\mu = \sum_{|\alpha|=n} \int_\Omega (D^\alpha f) u_\alpha dx \quad , \qquad f \in C^n(\bar{\Omega}) \tag{3.4}$$

where dx *is Lebesgue measure in* $\mathbb{R}^d$. *Moreover, for* $d > n$, *there exists* μ *satisfying the hypothesis but such that no formula of the form (3.4), with all* u_α *in* $L^{d/d-n}(\Omega)$, *is possible.*

Proof Let us first prove the latter assertion. Let ξ be some point in Ω . It is easy to show there exists a function $\phi \in L^\infty(\Omega)$ such that the measure $d\mu = \delta_{(\xi)} - \phi dx$ annihilates P_{n-1} . Suppose, for this choice of μ , (3.4) holds with all $u_\alpha \in L^p$ where $p = \frac{d}{d-n} < \infty$. Let us show that this leads to a contradiction. We have

$$|f(\xi) - \int f\phi dx| \le \sum_{|\alpha|=n} ||u_\alpha||_p \; ||D^\alpha f||_q$$

where $q = \frac{p}{p-1} = \frac{d}{n}$. This gives an estimate

$$|f(\xi)| \le A(\sum_{|\alpha|\le n} \int_\Omega |D^\alpha f|^q)^{1/q}$$

where A is a constant independent of f , valid for all smooth functions f . However, such an estimate is known to be false, indeed it would imply a *false* embedding theorem $W^{n,\frac{d}{n}} \to L^\infty$ in d dimensions (cf. Adams[1], p.118). We emphasize that we have assumed $n < d$.

For the other half of the theorem, assume now $p < \frac{d}{d-n}$ and let $q = \frac{p}{p-1}$. As earlier, let $(\alpha_1,\ldots,\alpha_m)$ be an enumeration of all multi-indices of rank n , and consider the map

$$T : f \to (f^{(\alpha_1)}, \ldots, f^{(\alpha_m)})$$

which we choose now, however, to regard as a map from $C^n(\bar{\Omega})$ to $L^q(\Omega)^m$. Let us define a linear functional λ on range T by

$$\lambda : (f^{(\alpha_1)}, \ldots, f^{(\alpha_m)}) \to \int f d\mu \quad .$$

This λ is well-defined, because if $Tf_1 = Tf_2$ then $f_1 - f_2$ is in P_{n-1} and hence $\int f_1 d\mu = \int f_2 d\mu$. We now claim that λ is *bounded* (where, recall, range T has the topology induced by $L^q(\Omega)^m$), that is, for some constant A

$$\left| \int f d\mu \right| \le A \sum_{|\alpha|=n} \| f^{(\alpha)} \|_q \quad . \tag{3.5}$$

Assuming this, λ extends to a bounded linear functional on $L^q(\Omega)^m$, which is representable in terms of a family $\{u_\alpha : |\alpha| = n\}$ of functions in $L^p(\Omega)$, and that gives (3.4).

So we have only to prove (3.5). First observe that, for any $g \in P_{n-1}$,

$$\left| \int f d\mu \right| = \left| \int (f-g) d\mu \right| \le \| f-g \|_\infty \cdot \| \mu \|_M \quad ,$$

hence (writing $A_1 = \| \mu \|_M$)

$$\left| \int f d\mu \right| \le A_1 \inf_{g \in P_{n-1}} \| f-g \|_\infty \quad . \tag{3.6}$$

We next invoke the Sobolev embedding theorem in the form

$$\| u \|_\infty \le A_2 [\| u \|_q + \sum_{|\alpha|=n} \| u^{(\alpha)} \|_q]$$

valid because $nq > d$ (cf. [1], p.97). We have conveniently omitted on the right terms involving derivatives of orders 1 through $n-1$, which is known to give an equivalent norm in $W^{n,q}(\Omega)$.

Thus

$$\|f-g\|_\infty \le A_2[\|f-g\|_q + \sum_{|\alpha|=n} \|f^{(\alpha)}\|_q]$$

$$\inf_{g\in P_{n-1}} \|f-g\|_\infty \le A_2[\inf_{g\in P_{n-1}} \|f-g\|_q + \sum_{|\alpha|=n} \|f^{(\alpha)}\|_q]$$

Comparing (3.5), (3.6) and the last estimate, we see that the theorem will be proved if we have an estimate

$$\inf_{g\in P_{n-1}} \|f-g\|_q \le A_3 \sum_{|\alpha|=n} \|f^{(\alpha)}\|_q \quad . \tag{3.7}$$

This is, however, a known inequality, a generalization of the *Poincaré inequality* (case $n=1$, $q=2$) due apparently to Sobolev[6] (see also Morrey[4], Sobolev[7]). This completes the proof.

Remark 1 If (3.7) is valid for $q = 1$, then in case $d = n$ the u_α in (3.4) can be chosen *bounded*.

Remark 2 As remarked earlier, we need not assume Ω convex: any domain for which (a) the Sobolev embedding theorem, and (b) the inequality (3.7), hold is enough. Concerning the latter condition, when $q > 1$ (3.7) can be deduced for any domain Ω for which the Rellich-Kondryashov compact embedding theorem holds. Geometric conditions sufficient for these theorems are given in the books[1,4].

Remark 3 One can extend Theorem 3 to the case $n > d$, by a similar (but technically more involved) argument. In this case the u_α have greater smoothness, which can be expressed by stating that they belong to certain Sobolov spaces $W^{k,p}(\mathbb{R}^d)$, $k = n-d$. One can also replace the hypothesis $R \in M(\bar{D})$ by other natural ones, e.g. membership in some Sobolev or Hölder class, and obtain representations (3.4) with u_α in corresponding (best possible) classes. For reasons of space we propose to return to these matters in a later paper, which will also deal with the more general situation when $\mathcal{P}_{n-1}$, the simultaneous nullspace of the associated differential operators $D^\alpha : |\alpha| = n$, is replaced by the nullspaces of other families of differential operators.

REFERENCES

1. Adams, R. (1975). Sobolev Spaces, Academic Press.
2. Davis, P. (1963). Interpolation and Approximation, Blaisdell.
3. Dunford, N. and Schwartz, J. (1964). Linear Operators, vol.I, Interscience (Second printing).
4. Morrey, C. (1966). Multiple Integrals in the Calculus of Variations, Springer.
5. Radon, J. (1935). Restausdrücke bei Interpolations- und Quadraturformeln durch bestimmte Integrale, *Monatsh. Math. Phys.* 42, 389-396.
6. Sobolev, S.L. (1963). Applications of Functional Analysis in Mathematical Physics, translated from the Russian, *AMS*.
7. Sobolev, S.L. (1974). Introduction to the Theory of Cubature Formulas, Moscow (Russian).
8. Sard, A. (1963). Linear Approximation, *AMS*.
9. Sard, A. (1965). Function Spaces, *Bull. A.M.S.* 71, 397-418.
10. Stroud, A. (1971). Approximate Calculations of Multiple Integrals, Prentice-Hall.

THE TRACE TO CLOSED SETS OF THE ZYGMUND CLASS OF QUASI-SMOOTH FUNCTIONS IN R^n

Hans Wallin

Institute of Mathematics and Statistics
University of Umeå
Umeå, Sweden

1. INTRODUCTION

The Zygmund class $\Lambda_1(R^n)$ of quasi-smooth functions in R^n is defined in the following way using the notation

$$\Delta_h^2 f(x) = f(x+h) - 2f(x) + f(x-h) \quad .$$

Definition 1 $f \in \Lambda_1(R^n)$ if f is continuous in R^n and, for some constant A ,

$$|f(x)| \le A \quad \text{and} \quad |\Delta_h^2 f(x)| \le A|h| \quad \text{for} \quad x,h \in R^n \quad .$$

The *norm* of f in $\Lambda_1(R^n)$ is the infimum of the constants A for which the inequalities hold.

Consider the following two problems where F , as always in this paper, is a closed subset of R^n .

Problem 1 Characterize the trace class

$$\{f|F: f \in \Lambda_1(R^n)\}$$

($f|F$ denotes the pointwise restriction of f to F).

Problem 2 Define a smoothness condition for functions defined only on F analogous to the condition on the second difference $\Delta_h^2 f$ in Definition 1. (If F is not convex you cannot automatically use second differences; the point x may belong to CF even if $x+h$ and $x-h$ belong to F.)

Reporting on joint research with A. Jonsson I shall give the solution to these two problems. In a slightly more general version, but with more complicated proofs, the solution is given in [1] which also contains further information on the problems including the rather obvious generalization to the case with higher differences $\Delta_h^k f(x)$. The analogous problems where the second difference is changed to the first difference were completely solved in 1934 by Whitney[3]. Whitney's study was motivated by the Dirichlet problem for the Laplace equation. Our motivation was the fact that in many problems second differences occur more naturally than first differences (see Zygmund[4]).

We first describe Whitney's result.

2. THE WHITNEY EXTENSION THEOREM

By definition, $f \in \mathrm{Lip}(\alpha,F)$, $0 < \alpha \le 1$, if for some constant A, $|f(x)| \le A$ and $|f(x)-f(y)| \le A|x-y|^\alpha$ for $x,y \in F$. The norm of f in $\mathrm{Lip}(\alpha,F)$ is the infimum of the constants A. It is comparatively easy to prove that every $f \in \mathrm{Lip}(\alpha,F)$ may be extended to R^n to a function in $\mathrm{Lip}(\alpha,R^n)$. The fact that this is true also for $\alpha > 1$ is the Whitney extension theorem. A major problem is to define $\mathrm{Lip}(\alpha,F)$, $\alpha > 1$. We shall consider only the case $1 < \alpha \le 2$ and refer to [2], p.176 for $\alpha > 2$. We use the following notation:

$$j = (j_1,\dots,j_n)\ ,\quad j_\nu \text{ integers, } \ge 0\ ;$$

$$|j| = j_1 + \dots + j_n\ ;\quad x^j = x_1^{j_1} \dots x_n^{j_n}$$

where $x = (x_1,\ldots,x_n) \in R^n$; D^j denotes the derivative $D_1^{j_1}\ldots D_n^{j_n}$.

Definition 2 $f \in \mathrm{Lip}(\alpha,F)$, $1<\alpha\le 2$, if f is defined on F , and there exist functions $\{f^{(j)}\}$, $|j|\le 1$, defined on F , with $f^{(0)} = f$ such that, for some constant A and all $x,y \in F$,

$$|f^{(j)}(x)| \le A \quad \text{for} \quad |j| \le 1 ,$$

$$|f(x)-f(y) - \sum_{|j|=1} (x,y)^j f^{(j)}(y)| \le A|x-y|^{\alpha} ,$$

$$|f^{(j)}(x) - f^{(j)}(y)| \le A|x-y|^{\alpha-1} \quad \text{for} \quad |j| = 1 .$$

We also say that $\{f^{(j)}\}_{|j|\le 1} \subset \mathrm{Lip}(\alpha,F)$. The *norm* of $\{f^{(j)}\}_{|j|\le 1} \in \mathrm{Lip}(\alpha,F)$ is the infimum of the constants A .

It is easy to see that $f \in \mathrm{Lip}(\alpha,R^n)$, $1<\alpha\le 2$, iff f is a bounded function in $C^1(R^n)$ and $D^j f \in \mathrm{Lip}(\alpha-1,R^n)$ for $|j| = 1$. In that case we also have $f^{(j)} = D^j f$ for $|j| = 1$.

THEOREM 1 (The Whitney extension theorem) *Every function* $f \in \mathrm{Lip}(\alpha,F)$ *may be extended to* R^n *to a function* $E_1 f \in \mathrm{Lip}(\alpha,R^n)$ *such that*

$$||E_1 f||_{\mathrm{Lip}(\alpha,R^n)} \le c||f||_{\mathrm{Lip}(\alpha,F)} .$$

Here and below c denotes a constant not depending on f and F . See [2], Ch. VI, Section 2, for a proof.

3. CHARACTERIZATION OF $\Lambda_1(R^n)$ BY APPROXIMATION. DEFINITION OF $\Lambda_1(F)$

We shall solve the Problems 1 and 2 by approximating $f \in \Lambda_1(R^n)$ by smoother functions.

Suppose that $f \in \Lambda_1(R^n)$. We take a function $\psi \in C_0^\infty(R^n)$ such that $\psi(x) = 0$ if $|x| \geq 1$, $\psi \geq 0$, $\psi(x) = \psi(-x)$, and $\int\psi dx = 1$. For $\nu = 1,2,\ldots$ we define ψ_ν by $\psi_\nu(x) = 2^{\nu n}\psi(2^\nu x)$ and f_ν by

$$f_\nu(x) = (f * \psi_\nu)(x) = \int f(x-t)\psi_\nu(t)dt = \int f(x+t)\psi_\nu(t)dt \quad .$$

Then, by the properties of ψ_ν and Definition 1,

$$|2(f_\nu(x)-f(x))| = |\int_{|t|\leq 2^{-\nu}} (f(x+t)+f(x-t)-2f(x))\psi_\nu(t)dt| \leq A2^{-\nu} \quad . \qquad (3.1)$$

Similarly, for $|j| = 2$,

$$|2D^j f_\nu(x)| = |\int_{|t|\leq 2^{-\nu}} \Delta_t^2 f(x) D^j\psi_\nu(t)dt| \leq cA2^\nu \quad .$$

This and the mean-value theorem give

$$|f_\nu(x)-f_\nu(y) - \sum_{|j|=1} (x-y)^j D^j f_\nu(y)| \leq cA|x-y|^2 \cdot 2^\nu , \qquad (3.2)$$

and

$$|D^j f_\nu(x)-D^j f_\nu(y)| \leq cA|x-y| \cdot 2^\nu \quad \text{if} \quad |j| = 1 \quad . \qquad (3.3)$$

Furthermore, it follows directly from the definition of f_ν that

$$|f_\nu(x)| \leq A \quad \text{and} \quad |D^j f_\nu(x)| \leq cA2^\nu \quad \text{for} \quad |j| = 1 \, . \qquad (3.4)$$

From (3.1) and (3.2) we also find by inserting certain terms

$$\Big|\sum_{|j|=1}(y-x)^j(D^jf_\nu(x)-D^jf_\mu(x))\Big| \le |f_\nu(y)-f_\mu(y)|$$
$$+\ |f_\mu(x)-f_\nu(x)|\ +\ |(f_\mu(y)-f_\nu(y))\ +\ (f_\nu(x)-f_\mu(x))$$
$$+\ \sum_{|j|=1}(y-x)^j(D^jf_\nu(x)-D^jf_\mu(x))| \le cA(2^{-\nu}+2^{-\mu})$$
$$+\ cA|x-y|^2(2^\nu+2^\mu) \quad .$$

By suitable choices of y, $|x-y| = 2^{-\nu}$, we conclude that

$$|D^jf_\nu(x)-D^jf_\mu(x)| \le cA \quad \text{if} \quad \nu = \mu+1\ ,\ |j| = 1\ . \tag{3.5}$$

Conversely, we shall prove that *if* $\{f_\nu\}$ *are functions satisfying* (3.1), (3.2) *and* (3.4), *then* $f \in \Lambda_1(R^n)$. In fact, when $|h|$ is small, choose ν such that $2^{-\nu} < |h| \le 2^{-\nu+1}$ and consider

$$\Delta_h^2 f(x) = (\Delta_h^2 f(x) - \Delta_h^2 f_\nu(x)) + \Delta_h^2 f_\nu(x) = I + II\ . \tag{3.6}$$

Then $|I| \le cA2^{-\nu} \le cAh$ by (3.1) and by inserting suitable terms we see by (3.2) that $|II| \le cA|h|^2 2^\nu \le cA|h|$. This proves the converse part.

The conclusion is that we may use the properties (3.1) - (3.5) to define a Zygmund class $\Lambda_1(F)$ of quasi-smooth functions on an arbitrary closed set F. Observe that, by Definition 2, (3.2) - (3.4) mean that $f_\nu \in \mathrm{Lip}(2,F)$ with norm $\le cA2^\nu$.

Definition 3 $f \in \Lambda_1(F)$ if F is defined on F and there exist functions $\{f_\nu^{(j)}\}$, $|j| \le 1$, $\nu = 1,2,\ldots,$ defined on F (we put $f_\nu^{(0)} = f_\nu$), such that, for some constant A and

and all $x \in F$, $\nu = 1,2,\ldots,$

(a) $|f_\nu(x)-f(x)| \leq A2^{-\nu}$

(b) $|f_{\nu+1}^{(j)}(x)-f_\nu^{(j)}(x)| \leq A$ for $|j| = 1$

(c) $\{f_\nu^{(j)}\}_{|j|\leq 1} \in \mathrm{Lip}(2,F)$ and $\|\{f_\nu^{(j)}\}\|_{\mathrm{Lip}(2,F)} \leq A^{2\nu}$.

The *norm* of f in $\Lambda_1(F)$ is the infimum of all A such that (a)-(c) hold for some $\{f_\nu^{(j)}\}$.

We note that f is bounded on F by (a), since each f_ν is bounded by (c). By (a) this means that f_ν is bounded uniformly in ν. The condition (b) follows from (a) and (c) for all "nice" sets F (compare the derivation of (3.5)) but not for all sets F (see [1], p.8). Definition 3 is based upon approximating f by functions $f_\nu \in \mathrm{Lip}(2,F)$ but it would also be possible to use functions in $\mathrm{Lip}(\alpha,F)$, $\alpha > 1$.

The discussion preceding Definition 3 gives

PROPOSITION 1 *The space* $\Lambda_1(F)$ *in Definition* 3 *coincides, when* $F = R^n$, *with the space* $\Lambda_1(R^n)$ *in Definition* 1, *and the norms introduced are equivalent.*

4. THE EXTENSION THEOREM AND THE EXTENSION OPERATOR

We now come to our main result.

THEOREM 2 (The extension theorem) *Every function* $f \in \Lambda_1(F)$ *may be extended to* R^n *to a function* $Ef \in \Lambda_1(R^n)$ *such that* $\|Ef\|_{\Lambda_1(R^n)} \leq c\|f\|_{\Lambda_1(F)}$, *where* c *is a constant depending only on* n.

Conversely, Proposition 1 implies that the restriction Rf to F of a function $f \in \Lambda_1(R^n)$ belongs to $\Lambda_1(F)$ and that $||Rf||_{\Lambda_1(F)} \leq c||f||_{\Lambda_1(R^n)}$.

The Problems 1 and 2 in Section 1 are clearly solved by means of the extension theorem, Proposition 1, and Definition 3. If F is the closure of a "Lipschitz domain", $\Lambda_1(F)$ may alternatively be characterized in the usual way by means of second differences. Together with Theorem 2 this leads to an alternative and well-known solution to the Problems 1 and 2 for "Lipschitz-domains" (see [1], Section 4, for details).

In order to describe the extension operator $f \to Ef$ we start from a function $f \in \Lambda_1(F)$ and a family $\{f_\nu^{(j)}\}$, $|j| \leq 1$, $\nu = 1,2,\ldots,$ satisfying the conditions in Definition 3; for simplicity in notation we normalize so that $A = 1$. By the Whitney extension theorem f_ν may be extended to a function $E_1 f_\nu \in \mathrm{Lip}(2,R^n)$ with norm $\leq c2^\nu$. This extension may be given by a formula of the following kind (see [2], p.177)

$$(E_1 f_\nu)(x) = \begin{cases} f_\nu(x), & x \in F \\ \sum_i{}' \phi_i(x)\{f_\nu(p_i) + \sum_{|j|=1} (x-p_i)^j f_\nu^{(j)}(p_i)\}, & x \in CF . \end{cases}$$

Here ϕ_i , $i = 1,2,\ldots,$ are certain non-negative functions in $C_0^\infty(R^n)$ satisfying $\Sigma\phi_i(x) = 1$, $x \in CF$, and

$$|D^j\phi_i(x)| \leq c_j(d(x,F))^{-|j|}$$

($d(x,F)$ is the distance from x to F) ; $p_i \in F$ are points such that $|x-p_i| \leq cd(x,F)$ for x belonging to the support of ϕ_i , supp ϕ_i ; the symbol Σ' indicates that the sum is taken only over those i such that $\mathrm{supp}\,\phi_i \subset \{x: d(x,F) \leq 1\}$.

The extension Ef is now defined by means of the functions $E_1 f_\nu$ and a partition of unity which we describe in the following lemma.

LEMMA 1 *Let* $\Delta_\nu = \{x: 2^{-\nu-1} < d(x,F) \le 2^{-\nu}\}$. *Then there exist non-negative functions* $\Phi_\nu \in C^\infty(R^n)$, $\nu = 0,\pm 1,\pm 2,\ldots,$ *such that* $\Phi_\nu(x) = 0$ *if* $x \notin \Delta_{\nu-1} \cup \Delta_\nu \cup \Delta_{\nu+1}$, $\Sigma\Phi_\nu(x) = 1$ *if* $x \in CF$, *and, for all* j , $|D^j\Phi_\nu(x)| \le c_j 2^{\nu|j|}$.

The functions Φ_ν are produced, roughly speaking, by convoluting certain C^∞-functions with support in $\{x: \; x \le 2^{-\nu-3}\}$ with the characteristic functions of certain sets which are close to the sets Φ_ν . For the details we refer to [1], Lemma 3.1.

Finally, we now define Ef by

$$(Ef)(x) = \begin{cases} f(x) , & x \in F \\ \sum_\nu \Phi_\nu(x)(E_1 f_\nu)(x) , & x \in CF \end{cases} .$$

5. PROOF OF THE EXTENSION THEOREM

We keep the assumptions from Section 4 and shall start by proving the following two properties:

$$|D^j(E_1 f_\nu)(x)| \le c2^\nu \quad \text{if} \quad x \in CF , \; |j| = 2 . \tag{5.1}$$

$$|E_1 f_\nu(x) - E_1 f_\mu(x)| \le c2^{-\mu} \text{ if } d(x,F) \le 2^{-\nu+1}, \; \nu \ge \mu - 1 . \tag{5.2}$$

The partial derivatives of $E_1 f_\nu$ of the first order are in $Lip(1,R^n)$ with norm $\le c2^\nu$, since $E_1 f_\nu \in Lip(2,R^n)$ with norm $\le c2^\nu$. Since $E_1 f_\nu \in C^\infty(CF)$, (5.1) now follows. In order to prove (5.2) we put

$$A_i(x) = (f_\nu(p_i)-f_\mu(p_i)) + \sum_{|j|=1} (x-p_i)^j (f_\nu^{(j)}(p_i)-f_\mu^{(j)}(p_i))$$

and note that by (a) in Definition 3, $|f_\nu(p_i)-f_\mu(p_i)| \le 2^{-\nu} + 2^{-\mu}$ and by repeated application of (b) in Definition 3, $|f_\nu^{(j)}(p_i)-f_\mu^{(j)}(p_i)| \le |\nu-\mu| \le 2^{|\nu-\mu|}$, if $|j| = 1$. These facts and the definition of $E_1 f_\nu$ imply for $x \in CF$

$$|E_1 f_\nu(x)-E_1 f_\mu(x)| = |\sum{}' \phi_i(x)A_i(x)| \le |\sum{}' \phi_i(x)\{(2^{-\nu}+2^{-\mu}) + cd(x,F) \cdot 2^{|\nu-\mu|}\}|$$

which gives (5.2). The last equation and the properties of ϕ_i also give, if $|j| = 1$,

$$|D^j(E_1 f_\nu)(x)-D^j(E_1 f_\mu)(x)| = |\sum_i{}' D^j\phi_i(x)\cdot A_i(x) + \sum_i{}' \phi_i(x)(f_\nu^{(j)}(p_i)-f_\mu^{(j)}(p_i))| \le c \quad, \tag{5.3}$$

if $|\nu-\mu| \le 1$ and $x \in \Delta_\mu$, i.e. $2^{-\mu-1} < d(x,F) \le 2^{-\mu}$.

Using these properties we next prove two lemmas.

LEMMA 2 $|D^j(Ef)(x)| \le c(d(x,F))^{-1}$ *if* $x \in CF$, $|j| = 2$.

Proof Let $|j| = 2$ and suppose that $x \in \Delta_\mu$. $D^j(Ef)(x)$ is a sum of terms of type

$$\sum_{\nu=\mu-1}^{\mu+1} D^{j-\ell}\Phi_\nu(x) \cdot D^\ell(E_1 f_\nu)(x) \quad, \quad |\ell| \le 2 \quad.$$

For $|\ell| = 2$ these terms are estimated by means of (5.1) and we get an upper bound $c2^\mu \le c(d(x,F))^{-1}$. For $|\ell| = 1$ we get by means of $\{\Phi_\nu\}$ and (5.3)

$$|\sum D^{j-\ell}\Phi_\nu(x)D^\ell(E_1 f_\nu)(x)| = |\sum D^{j-\ell}\Phi_\nu(x)(D^\ell(E_1 f_\nu)(x)$$

$$- D^\ell(E_1 f_\mu)(x))| \le c2^\mu \le c(d(x,F))^{-1} \quad .$$

For $\ell = 0$ we get analogously by (5.2)

$$|\sum D^j\Phi_\nu(x)E_1 f_\nu(x)| = |\sum D^j\Phi_\nu(x)(E_1 f_\nu(x)-E_1 f_\mu(x))| \le c2^{\mu|j|}2^{-\mu}$$

$$= c2^\mu \le c(d(x,F))^{-1} \quad .$$

Taken together these three estimates prove the lemma.

LEMMA 3 $|(Ef)(x)-(E_1 f_\mu)(x)| \le c2^{-\mu}$ *if* $d(x,F) \le 2^{-\mu-1}$.

Proof When $x \in F$ this is true by (a) in Definition 3. Now, let $x \in CF$, say $x \in \Delta_{\nu_0}$, i.e. $2^{-\nu_0-1} < d(x,F) \le 2^{-\nu_0}$. Since $d(x,F) \le 2^{-\mu-1}$, $\nu_0 > \mu$. Hence, by (5.2),

$$|Ef(x)-E_1 f_\mu(x)| = |\sum_{\nu_0-1}^{\nu_0+1} \Phi_\nu(x)(E_1 f_\nu(x)-E_1 f_\mu(x))| \le c2^{-\mu} \quad .$$

We now turn to the proof of Theorem 2.

Proof that $|Ef(x)| \le c$. We know that $|f_\nu| \le c$ on F and $|f_\nu^{(j)}| \le c2^\nu$ on F , if $|j| = 1$. Hence, by the definition of $E_1 f_\nu$, $|E_1 f_\nu(x)| \le c$ if $d(x,F) \le 2^{-\nu+1}$. This and the definition of Ef give that $|Ef(x)| \le c$ for all x since $\Phi_\nu(x) = 0$ if $x \notin \Delta_{\nu-1} \cup \Delta_\nu \cup \Delta_{\nu+1}$.

Proof that $|\Delta_h^2(Ef)(x)| \le c|h|$. We consider two cases. Case 1: $d(x,F) \ge 2|h| > 0$. In this case the line segment L between $x-h$ and $x+h$ does not intersect F . Since $Ef \in C^\infty(CF)$ the mean-value theorem and Lemma 2 give, if

$d(L,F)$ denotes the distance from L to F,

$$|\Delta_h^2(Ef)(x)| \leq c|h|^2(d(L,F))^{-1} \leq c|h| \quad .$$

Case 2: $d(x,F) < 2|h| \neq 0$. We may assume that h is so small that we can choose an integer $\mu \geq 1$ such that $2^{-\mu-2} < d(x,F) + |h| \leq 2^{-\mu-1}$. Consider

$$\Delta_h^2 Ef(x) = (\Delta_h^2 Ef(x) - \Delta_h^2 E_1 f_\mu(x)) + \Delta_h^2 E_1 f_\mu(x) = I + II \quad .$$

By Lemma 3, $|I| \leq c2^{-\mu} \leq c|h|$. As in the estimation of II in (3.6), we find that $|II| \leq c2^{\mu}|h|^2 \quad c|h|$, and hence $|\Delta_h^2(Ef)(x)| \leq c|h|$.

Proof that Ef *is continuous* Inserting the continuous function $E_1 f_\mu$ it follows from Lemma 3 that Ef is continuous at $x \in F$ and hence everywhere. By checking with Definition 1 we see that Theorem 2 is proved.

REFERENCES

1. Jonsson, A. and Wallin, H. (1977). The Trace to Closed Sets of Functions with Second Difference of Order O(h), *Dept. of Math., Univ. of Umeå* 7.

2. Stein, E.M. (1970). Singular Integrals and Differentiability Properties of Functions, Princeton, Princeton Univ. Press.

3. Whitney, H. (1934). Analytic Extensions of Differentiable Functions Defined in Closed Sets, *Trans. Amer. Math. Soc.* 36, 63-89.

4. Zygmund, A. (1945). Smooth Functions, *Duke Math. J.* 12, 47-76.

APPROXIMANTS FOR SOME CLASSES OF MULTIVARIABLE FUNCTIONS PROVIDED BY VARIATIONAL PRINCIPLES

Michael F. Barnsley

Service de Physique Atomique
Centre d'Etudes Nucléaires de Saclay
Gif Sur Yvette, France

1. INTRODUCTION

By using a specific example we illustrate how variational principles can sometimes be used to discover multivariable approximants appropriate for various classes of functions. Such an approach has previously been used to obtain bounds for a variety of problems involving one-variable functions, see for example[1,2,3].

We consider explicitly two-variable Stieltjes functions expressible in the form

$$f(w,z) = \int_0^\infty \int_0^\infty \frac{d\sigma(s,t)}{(1+ws+zt)} \quad ; \tag{1.1}$$

where $\sigma(s,t)$ on $0 \le s < \infty$, $0 \le t < \infty$, is bounded, monotone nondecreasing in t for fixed s , and nondecreasing in s for fixed t ; and where the variables w and z both belong to the complex plane cut from $-\infty$ to $0-$. This function has the formal double-series expansion

$$F(w,z) \sim \sum_{m=0}^{\infty} \sum_{n=0}^{\infty} (-1)^{m+n} \binom{m+n}{n} F_{m,n} w^m z^n \tag{1.2}$$

where

$$F_{m,n} = \int_0^\infty \int_0^\infty s^m t^n \, d\sigma(s,t) \qquad m,n = 0,1,\ldots \tag{1.3}$$

We suppose that we know an initial set of coefficients occurring in (1.2), say

$$F_{0,0}, F_{0,1}, F_{0,1}, F_{2,0}, F_{1,1}, F_{0,2}, \ldots, F_{j,k} \quad , \tag{1.4}$$

and assume that these are finite. Then we ask what bounds can be imposed on $F(w,z)$ when $0 \le w < \infty$, $0 \le z < \infty$, on the basis of the given information.

This problem is of particular interest because (1.1) is a possible generalization of single-variable Stieltjes functions: our results provide a possible definition for two-variable rational approximants such that their relation with (1.1) is analogous to the relationship between Padé approximants and one-variable Stieltjes functions; see also [4].

In §2 we provide complementary variational functionals $J(\Phi)$ and $G(\Psi)$ such that

$$J(\Phi) \le F(w,z) \le G(\Psi) \quad , \qquad w \ge 0 \, , \; z \ge 0 \, , \tag{1.5}$$

for all Φ and Ψ lying within a suitable Hilbert space domain. By making appropriate choices for the trial vectors we find rational approximants $J(N,R,S)$ and $G(N,R,S)$ with

$$J(N,R,S) \le F(w,z) \le G(N,R,S) \quad \text{whenever} \quad w \ge 0 \quad \text{and} \quad z \ge 0 \, , \tag{1.6}$$

where (N,R,S) indexes the $F_{m,n}$'s which are used. The properties of these approximants are briefly described. Further details can be found in [5].

In §3 we discuss some other situations where the present

approach may be applied.

2. DUAL VARIATIONAL BOUNDS AND RATIONAL APPROXIMANTS FOR F(w,z)

We begin as in [6] by writing

$$F(w,z) = \langle f,(1+wA+zB)^{-1}f\rangle \quad , \quad w \geq 0 \ , \ z \geq 0 \ , \tag{2.1}$$

where $\langle\cdot,\cdot\rangle$ denotes the inner product in a real Hilbert space h , $f \in h$, and A and B are a commuting pair of positive self-adjoint operators in h with domains D(A) and D(B) respectively. Specifically, h is the Hilbert space of real functions $h(s,t)$, $0 \leq s < \infty$, $0 \leq t < \infty$, which are square integrable with respect to the measure $\sigma(s,t)$, and the inner product between h_1 and h_2 in h is

$$\langle h_1,h_2\rangle = \int_0^\infty \int_0^\infty h_1(s,t)h_2(s,t)d\sigma(s,t) \quad . \tag{2.2}$$

A and B are the linear operators which multiply by s and t respectively, so that for example

$$A\eta = A\eta(s,t) = s\eta(s,t) = \tilde{\eta} \quad \text{for all} \ \ \eta \in D(A) \quad , \tag{2.3}$$

where D(A) is precisely the set of $\eta \in h$ such that $\tilde{\eta} \in h$. Then, since both A and B are both positive, the equation

$$(I+L)\phi = f, \ \phi \in D(L) = D(A) \cap D(B) \ , \quad \text{where} \quad L = wa + zB \ , \tag{2.4}$$

has a unique solution ϕ for each $f \in h$ when $w \geq 0, \ z \geq 0$.

With

$$f = f(s,t) = 1 \quad \text{for all} \ \ 0 \leq s < \infty \ , \ 0 \leq t < \infty \ , \tag{2.5}$$

which belongs to h by virtue of the boundedness of $\sigma(s,t)$, we have in particular

$$<\phi,f> = <f,(I+L)^{-1}f> = \int_0^\infty \int_0^\infty \frac{d\sigma(s,t)}{(I+ws+zt)} = F(w,z) \quad . \tag{2.6}$$

Observe that the coefficients $F_{m,n}$ in (1.3) can be expressed

$$F_{m,n} = <f,A^m B^n f> \quad , \qquad m,n = 0,1,2,\ldots \tag{2.7}$$

On applying the theory of dual variational principles[7,8] to (2.4) one obtains the seemingly most elementary pair of complementary functionals, see [2] for example,

$$\begin{aligned} J(\Phi) &= -<\Phi,(I+L)\Phi> + 2<\Phi,f>, \\ G(\Psi) &= <f,f> + <\Psi,L(1+L)\Psi> - 2<L\Psi,f> \quad , \end{aligned} \tag{2.8}$$

where Φ and Ψ belong to $D(L)$. These functionals constitute variational approximations to $<\Phi,f>$, and moreover they impose the bounds (1.5) which are valid for $w \geq 0,\ z \geq 0$. The common stationary value $<\Phi,f>$ is achieved when $\Phi = \Psi = \phi$.

If we choose trial vectors

$$\Phi,\Psi = \sum_{n=1}^{M} \alpha_n \theta_n \quad , \qquad \alpha_n \in \mathbb{R}\ , \ \ \theta_n \in D(L) \tag{2.9}$$

and then maximise $J(\Phi)$ and minimise $G(\Psi)$ with respect to the α_n's , the result is bounds on $F(w,z)$ in the form of rational functions. Explicitly

$$J(\Phi_{opt}) = \underline{a}^+ \underline{\underline{b}}\underline{a} \quad , \quad \text{and} \quad G(\Psi_{opt}) = <f,f> - \underline{c}^+ \underline{\underline{d}}^{-1} \underline{c} \quad , \tag{2.10}$$

where $\underline{a}$ and $\underline{c}$ are column vectors with elements $<\theta_n,f>$ and $<\theta_n,Lf>$ $(n = 1,2,\ldots,M)$, respectively, $\underline{\underline{b}}$ and $\underline{\underline{d}}$ are matrices with elements $<\theta_n,(I+L)\theta_m>$ and $<\theta_n,L(I+L)\theta_m>$ $(m,n = 1,2,\ldots,M)$, respectively, and where + denotes transposition.

In order that the approximants (2.10) should require for their construction only an 'initial' set of $F_{m,n}$'s occurring in the expansion (1.2), we take as basis set all of the vectors in (α) and (β) below:

(α) f, together with all vectors of the form $A^m B^n f$ where $m+n = 1,2,\ldots,N$; m,n and N being non-negative integers, with N given.

(β) $A^R B^{N+1-R} f, A^{R-1} B^{N+2-R} f, \ldots, A^{N+1-S} B^S f$ where R and S are given integers such that $R+S \geq N$, $0 \leq R \leq N+1$, $0 \leq S \leq N+1$. When $R+S = N$ we understand that there are no vectors in (β).

This basis set can be indexed by the triple (N,R,S), and the total number of vectors is $M = \frac{1}{2}(N+1)(N+2) + (R+S-N)$. Assuming they have been ordered we denote them by $\theta_1, \theta_2, \ldots, \theta_M$ just as in (2.9). The corresponding pair of approximants in (2.10) are denoted by $J(N,R,S)$ and $G(N,R,S)$ respectively. Using the relation (2.7) and the fact that A and B commute we find that in order to explicitly construct $J(N,R,S)$ we need to know the coefficients

$$\begin{aligned}
&\{F_{m,n} : m+n = 0,1,\ldots,2N+1,\ m \geq 0,\ n \geq 0\} \\
&\{F_{N+1+R,N+1-R}, F_{N+R,N+2-R}, \ldots, F_{N+1-S,N+1+S}\} \qquad (2.11) \\
&\{F_{2R+1,2N+2-2R}, F_{2R,2N+3-R}, \ldots, F_{2N+2-2S,2S+1}\}
\end{aligned}$$

where the latter two sets must be taken to be empty when $R+S = N$. A similar looking set of coefficients is needed for the construction of $G(N,R,S)$. By adjusting (N,R,S) one can ensure maximal usage of a given set of coefficients such as (1.4). Two simple approximants are:

$$J(1,0,0) = -\frac{\begin{vmatrix} 0 & F_{0,0} & F_{1,0} & F_{0,1} \\ F_{0,0} & (F_{0,0}+wF_{1,0}+zF_{0,1}) & (F_{1,0}+wF_{2,0}+zF_{1,1}) & (F_{0,1}+wF_{1,1}+zF_{0,2}) \\ F_{1,0} & (F_{1,0}+wF_{2,0}+zF_{1,1}) & (F_{2,0}+wF_{3,0}+zF_{2,1}) & (F_{1,1}+wF_{2,1}+zF_{1,2}) \\ F_{0,1} & (F_{0,1}+wF_{1,1}+zF_{0,2}) & (F_{1,1}+wF_{2,1}+zF_{1,2}) & (F_{0,2}+wF_{1,2}+zF_{0,3}) \end{vmatrix}}{\text{(Same determinant without first row and column)}} \tag{2.12}$$

and

$$G(0,0,0) = F_{0,0} - \frac{(wF_{1,0}+zF_{0,1})^2}{(wF_{1,0}+zF_{0,1}+w^2F_{2,0}+2wzF_{1,1}+z^2F_{0,2})} \tag{2.13}$$

It follows from the variationally optimal nature of the bounds (1.6) that they must improve as the basis set is enlarged. Thus we obtain, for example, for allowed (N,R,S) and $(N,R+1,S)$,

$$J(N,R,S) \le J(N,R+1,S) \le J(N+1,R,S) \ldots \le F(w,z)$$

and

$$F(w,z) \le \ldots \le G(N+1,R,S) \le G(N,R+1,S) \le G(N,R,S) , \tag{2.14}$$

for all $w \ge 0$ and $z \ge 0$. It is also possible to prove, using the variational characterization of these approximants, that as w and z tend to zero

$$F(w,z) - J(N,R,S) \sim \text{terms of order } w^p z^q,\ p+q \geq 2N+2\ ,$$

and

$$F(w,z) - G(N,R,S) \sim \text{terms of order } w^p z^q,\ p+q \geq 2N+3\ , \tag{2.15}$$

which are analogous to the 'agreement-through-order' properties of one-variable Padé approximants, see [9]. The proof of this result can be found in [5]. Approximants of the type $J(N,0,0)$ are also considered in [6].

Example: we take

$$F(W,Z) = \int_1^2 ds \int_0^\infty \frac{dt\ \exp(-t^2)}{(1+ws+zt)} \tag{2.16}$$

which is of the form (1.1). Then at $w = z = 1$ we find

$$J(1,1,0) = 0.2969 \leq F(1,1) \leq 0.3089 = G(0,0,1)\ . \tag{2.17}$$

3. SOME OTHER SITUATIONS WHERE VARIATIONAL PRINCIPLES CAN PROVIDE MULTIVARIABLE APPROXIMANTS

A similar treatment to the one outlined above can clearly be carried through for functions expressible in the form

$$H(w,z) = \int_0^\infty \int_0^\infty \frac{d\sigma(s,t)}{(1+ws)(1+zt)}\ , \quad w \geq 0\ , \quad z \geq 0\ . \tag{3.1}$$

The formulation is essentially the same as before except that one must now choose $L = wA+zB+wzAB$. The resulting approximants are again of rational structure, but are distinct from $J(N,R,S)$ and $G(N,R,S)$. They appear to be closely related to Chisholm approximants [4].

More generally, for any function $K(w,z)$ known to be analytic and regular for all (w,z) lying within and on $C_1 \times C_2$ where C_1 and C_2 are smooth closed complex contours which enclose the origin we have

$$K(w,z) = -\frac{1}{4\pi^2}\int_{C_1}\int_{C_2}\frac{K(\xi,\eta)d\xi d\eta}{(\xi-w)(\eta-z)} \quad , \tag{3.2}$$

for all (w,z) interior to $C_1 \times C_2$. Following the lines laid out in [3] it is possible to formulate a Hilbert space together with linear operators P and Q , and vectors g_1 and g_2 such that

$$K(w,z) = \langle\phi, g_1\rangle \quad , \tag{3.3}$$

where ϕ is the unique solution of

$$(P-w)(Q-z)\phi = g_2 \quad . \tag{3.4}$$

Then the functional

$$J(\Psi,\Phi) = -\langle\Phi,(Q-z)^*(P-w)^*\Psi\rangle + \langle\Phi,g_1\rangle + \langle g_2,\Psi\rangle \quad , \tag{3.5}$$

where '*' denotes 'adjoint' and Φ and Ψ belong to appropriate domains, constitutes a bivariational approximation to $K(w,z)$, see [11] . That is,

$$K(w,z) - J(\Psi,\Phi) = \langle\delta\Phi,(Q-z)^*(P-w)^*\delta\Psi\rangle \quad , \tag{3.6}$$

where $\delta\Phi = \Phi-\phi$ and $\delta\Psi = \Psi-\psi$, ψ being the solution of

$$(Q-z)^*(P-w)^*\psi = g_1 \quad . \tag{3.7}$$

By making $J(\Psi,\Phi)$ stationary with respect to variations in Φ and Ψ , when the latter are constrained to lie in suitably chosen subspaces, one discovers approximants which require for their construction an 'initial' set of coefficients occurring in the double series expansion of $K(w,z)$ about $w,z = 0$. An analogous procedure in the one variable case provides the usual Padé approximants, see [3].

One interesting class of two-variable functions which might be explored with the aid of a variational principle arises as follows. Let S, T and U, be bounded self-adjoint linear operators in a Hilbert-space h. Let $L(w,z)$ denote the lowest eigenvalue of the operator $(U+wS+zT)$ when w and z are real. Then we have the well-known upper bound

$$L(w,z) \leq \frac{\langle\Phi,(U+wS+zT)\Phi\rangle}{\langle\Phi,\Phi\rangle} \quad \text{for all} \quad \Phi \subset h \quad . \tag{3.8}$$

Write the formal expansion of $L(w,z)$ about $(0,0)$ in the form $\sum_{m=0}^{\infty}\sum_{n=0}^{\infty} L_{m,n}w^m z^n$, and let $\phi(w,z)$ denote an eigenfunction corresponding to $L(w,z)$. Then on choosing

$$\Phi = \phi(0,0) \tag{3.9}$$

we find

$$L(w,z) \leq L_{0,0} + wL_{1,0} + zL_{0,1} \quad \text{for all} \quad w,z \in \mathbb{R} \quad . \tag{3.10}$$

Can more elaborate choices of trial function be made to yield higher order approximants? In the one variable case they certainly can, see [10].

REFERENCES

1. Epstein, S.T. and Barnsley, M.F. (1973). A variational approach to the theory of multipoint Padé approximants, *J. Math. Phys.* 14, 314-325.
2. Barnsley, M.F. and Robinson, P.D. (1974). Dual variational principles and Padé type approximants, *J. Inst.*

Math. Appl. 14, 229-249.

3. Barnsley, M.F. and Baker, G.A. Jr. (1976). Bivariational bounds in a complex Hilbert space and correction terms for Padé approximants, *J. Math. Phys.* 17, 1019-1027.

4. Chisholm, J.S.R. (1973). Rational approximants defined from double series, *Math. Comp.* 27, 841-848.

5. Barnsley, M.F. and Robinson, P.D. (1977). Rational approximant bounds for a class of two-variable stieltjes functions, *S.I.A.M. (Math. Anal.)*, in press.

6. Alabiso, C. and Butera, P. (1975). N-variable rational approximants and method of moments, *J. Math. Phys.* 16, 840-845.

7. Noble, B. and Sewell, M.J. (1972). On dual extremum principles in applied mathematics, *J. Inst. Math. Appl.* 9, 123-193.

8. Arthurs, A.M. (1970). Complementary Variational Principles. Clarendon Press, Oxford.

9. Baker, G.A. Jr. (1965). The theory and application of the Padé approximant method, *Advan. Theor. Phys.* 1, 1-57.

10. Barnsley, M.F. and Aguilar, J. (1977). On the approximation of potential energy functions for two center systems, in preparation.

11. Barnsley, M.F. and Robinson, P.D. (1977). Bivariational bounds for nonlinear problems, *J. Inst. Math. Appl.* 10, in press.

SYMMETRY IN BIRKHOFF MATRICES

G.G. Lorentz

Department of Mathematics
University of Texas
Austin, Texas, USA

1. INTRODUCTION

A *Birkhoff interpolation problem* is the problem to find a polynomial P of degree at most n which satisfies conditions

$$P^{(k)}(x_i) = c_{ik} \quad , \quad e_{ik} = 1 \quad , \tag{1.1}$$

where $X: x_1 < x_2 < \ldots < x_m$ are given knots, c_{ik} given constants, and the $m \times (n+1)$ matrix of zeros and ones $E = (e_{ik})_{i=1}^{m} {}_{k=0}^{n}$ decides when an equation (1.1) is present: this happens if and only if $e_{ik} = 1$. Thus, there are $N = |E|$ equations (1.1), $|E|$ being the number of ones in the matrix E. The pair E,X is called *regular* if (1.1) is solvable for each selection of constants c_{ik} (and otherwise *singular*); this can happen only if $N \leq n+1$. The theory of linear equations tells that under this assumption, the pair E,X is regular exactly when the corresponding homogeneous equations have only a trivial solution $P \equiv 0$. In other words, a polynomial P (of degree $\leq n$) *annihilated by* E,X: $P \perp E,X$ must be identically zero. Then we must have $N = n+1$; in this case the matrix E is called *normal*; this we assume from now on.

A matrix E is *regular*, if the pair E,X is regular for all selections of the knots X. The so-called Pólya condition is necessary for the regularity of the matrix E. Even more: if it is violated, then E,X *is singular for all selections of knots* X (Nemeth[5], D. Ferguson[1]). The Pólya condition has the form

$$M_k \geq k+1 \quad , \qquad k = 0,\ldots,n \quad , \tag{1.2}$$

where M_k is the number of ones in the first $k+1$ columns of E. Of course, for a normal matrix, $M_n = n+1$. A *sequence* is a continuous block of ones in a row, which begins after a zero

$$0 \ \underbrace{1 \ 1 \ 1 \ldots 1}\ 0$$

or in a column numbered 0 and ends before the next zero, or in the last column n of E. An *odd* sequence has an odd number of ones. A sequence is *supported* if its first one $e_{ik} = 1$ is 'supported' by two other ones, $e_{i_1,k_1} = 1$, $e_{i_2,k_2} = 1$ in positions $i_1 < i$, $k_1 < k$; $i_2 > i$, $k_2 > k$.

There is only one known sufficient (but not necessary!) condition for regularity of a matrix. According to a theorem of Atkinson and Sharma, E is regular if it satisfies the Pólya condition, and has no odd supported sequences.

In contrast, there are several non-trivial conditions for singularity of E, obtained by this author. They are:

(α) E is singular, if one of its rows contains exactly one odd supported sequence (all other sequences of the row being even or unsupported) (see Lorentz[2]),

(β) E is singular, if it contains at least one odd supported sequence, and if all of its rows except at most one consist of exactly one one (see Lorentz[3]).

The following depends on the notion of coalescence, due to Karlin and Karon:

(γ) E is singular, if a certain combinatorial integer σ, which depends on multiple coalescences of rows of E, is odd (see Lorentz[4]).

In the present paper we offer some additional criteria of singularity, based upon symmetry properties of E. (Compare also the paper of Windhauer[6].)

2. SYMMETRIC MATRICES

A symmetric matrix E will always have an *odd number* $2m+1$ of rows; to assure full generality, we allow the central row $i = 0$ to be empty (to consist of zeros). In this case, the rows will be numbered $-m,\dots,0,\dots,m$; E is symmetric if $e_{-i,k} = e_{ik}$ for all $i \neq 0$, and all k.

A set of knots $X: x_{-m} < \dots < x_0 < \dots < x_m$ is symmetric if $x_{-i} = -x_i$, $i \neq 0$, $x_0 = 0$. In this situation, we denote by X_0 the set of knots $x_1,\dots,x_m$. By E_1 we denote the $(m+1) \times (n+1)$ matrix which has rows numbered $0,1,\dots,m$, its rows $1,\dots,m$ being the same as in E, while the row 0 of E_1 has ones at exactly ones of row 0 of E in even positions. Similarly, E_2 consists of rows $1,\dots,m$ of E augmented by row 0 which retains ones of row 0 of E in odd positions. Finally, $\bar{E}_1$ (or $\bar{E}_2$) are obtained from E_1 (or E_2) by replacing all elements in odd (or even) positions of row 0 by ones.

For a polynomial P of degree at most n, we denote by P_e (or P_o) its even (or odd) part:

$$P_e(x) = \tfrac{1}{2}\{P(x) + P(-x)\}\ , \qquad P_o(x) = \tfrac{1}{2}\{P(x) - P(-x)\}\ .$$

LEMMA. *Let* E, X *be a symmetric pair. A polynomial* P *is annihilated by* E, X *if and only if its even part* P_e *is annihilated by* E_1, X_0 *(or, equivalently, by* $\bar{E}_1, X_0$ *) and its odd part* P_o *is annihilated by* E_2, X_0 *(or, equivalently, by* $\bar{E}_2, X_0$ *).*

This follows at once from the formulas, valid for $e_{ik} = 1$,

$$P_e^{(k)}(x_i) = \tfrac{1}{2}\{P^{(k)}(x_i) + P^{(k)}(x_{-i})\}\ , \quad \text{if } k \text{ is even} \tag{2.1}$$

and

$$P_e^{(k)}(x_i) = \tfrac{1}{2}\{P^{(k)}(x_i) - P^{(k)}(x_{-i})\}\ , \quad \text{of } k \text{ is odd,} \tag{2.2}$$

and from similar formulas for P_o .

We call a symmetric matrix E with $n+1$ ones, $|E| = n+1$, *symmetrically regular*, if each polynomial P annihilated by the pair E, X with a symmetric X is identically zero. We can reduce symmetric regularity to ordinary regularity of a shorter matrix.

THEOREM A. *A symmetric matrix* E *is symmetrically regular if and only if both* $\bar{E}_1, \bar{E}_2$ *are regular. In addition, we must have*

$$p_e = p_o = q \quad , \tag{2.3}$$

where p_e *(or* p_o *) is the number of zeros of row* 0 *in* E *in even (or odd) positions, and* q *is the number of ones in rows* $1, \ldots, m$ *of* E .

Proof. (a) Let E be symmetrically regular. We wish to show that $\bar{E}_1$ is regular. Let P be a polynomial with $P \perp \bar{E}_1, X_0$.

Let X be the corresponding symmetric set of knots. Then P is even and $P \perp E,X$, hence $P \equiv 0$. In addition, we have to show that $\bar{E}_1$ is normal, $|\bar{E}_1| = n+1$. If $|\bar{E}_1| < n+1$, then the assumption $P \perp \bar{E}_1, X_0$ is equivalent to a system of less than $n+1$ homogeneous linear equations for the $n+1$ coefficients of P. Such a system always has a non-trivial solution. Since this is not the case, we have $|\bar{E}_1| \geq n+1$. Similarly we show $|\bar{E}_2| \geq n+1$. But

$$|\bar{E}_1| = q + (n+1) - p_e \quad , \quad |\bar{E}_2| = q + (n+1) - p_o \quad , \tag{2.4}$$

$$|E| = 2q + (n+1) - p_e - p_o = n + 1 \quad . \tag{2.5}$$

We have $2q = p_e + p_o$, $q \geq p_e$, $q \geq p_o$, hence equations (2.3). In addition, $|\bar{E}_1| = n+1$. This shows that $\bar{E}_1$ is regular. Similarly, $\bar{E}_2$ is regular.

(b) Let $\bar{E}_1$ and $\bar{E}_2$ be regular. Then $|\bar{E}_1| = |\bar{E}_2| = n+1$, and we again have (2.3). Let X be a symmetric set of knots, assume that $P \perp E,X$. According to the lemma, $P_e \mid \bar{E}_1, X_0$, hence $P_e \equiv 0$. Similarly $P_o \equiv 0$, so that $P = P_e + P_o \equiv 0$. This completes the proof.

Other necessary conditions for symmetric regularity are obtained from the fact that $\bar{E}_1, \bar{E}_2$ must satisfy the Pólya condition. Let q_k, p_{ek}, p_{ok} have the same meaning as q, p_e, p_o, but restricted to the first $k+1$ columns of E. Then we have

$$M_k = 2q_k + (k+1) - p_{ek} - p_{ok} \quad ,$$

and the Pólya condition for E becomes

$$p_{ek} + p_{ok} \leq 2q_k \quad , \tag{2.6}$$

while Pólya conditions for $\bar{E}_1, \bar{E}_2$ are, respectively,

$$p_{ek} \leq q_k \quad , \quad p_{ok} \leq q_k \quad , \tag{2.7}$$

with equalities for $k = n$. Note that conditions (2.6) and (2.7) together imply (2.3).

COROLLARY 1. *For a symmetric matrix* E , *conditions (2.3) and (2.7) are necessary for regularity.*

COROLLARY 2. *Each* 3 *row symmetric matrix* E *which satisfies (2.7) is symmetrically regular.*

Indeed, $\bar{E}_1, \bar{E}_2$ satisfy the Pólya condition, and each two row matrix satisfying it is regular.

COROLLARY 3. (Atkinson-Sharma theorem for symmetric regularity.) *A symmetric matrix* E *that satisfies (2.7) is symmetrically regular, if it contains no odd supported sequences in rows* $i \neq 0$.

Indeed, in this case $\bar{E}_1$ for example satisfies the Pólya condition and has no odd supported sequences. For any such sequence would begin with a one $e_{ik} = 1$, $i > 0$ supported on the right by some $e_{i',k'} = 1$, $i' > i$, $k' < k$, and then it would be supported in E on the left by $e_{-i',k'} = 1$.

EXAMPLES

1. Consider the matrices

$$E' = \begin{pmatrix} 1 & 1 & 1 & 0 & 0 & 0 & 0 \\ 0 & 1 & 0 & 1 & 0 & 1 & 0 \\ 1 & 0 & 0 & 0 & 0 & 0 & 0 \end{pmatrix}, \quad E'' = \begin{pmatrix} 1 & 1 & 0 & 0 & 0 & 0 \\ 1 & 0 & 1 & 0 & 0 & 0 \\ 1 & 1 & 0 & 0 & 0 & 0 \end{pmatrix}, \quad E''' = \begin{pmatrix} 1 & 1 & 0 & 0 & 0 & 0 \\ 1 & 0 & 0 & 1 & 0 & 0 \\ 1 & 1 & 0 & 0 & 0 & 0 \end{pmatrix}$$

E' is symmetrically singular, for $p_e = 4$, $p_o = 0$, hence it is singular. But this fact does not follow from Theorem α of Section 1. The matrix E'' is symmetrically singular for $p_e = 1$, $p_o = 3$, and also singular by Theorem α. As for E''', it is symmetrically regular by Corollary 2, yet singular by the deeper Theorem α.

2. The matrices of Túran's lacunary interpolation of the type $(0,2)$,

$$E' = \begin{pmatrix} 1 & 0 & 1 & 0 & 0 & 0 \\ 1 & 0 & 1 & 0 & 0 & 0 \\ 1 & 0 & 1 & 0 & 0 & 0 \end{pmatrix}, \qquad E'' = \begin{pmatrix} 1 & 0 & 1 & 0 & 0 & 0 & 0 & 0 \\ 1 & 0 & 1 & 0 & 0 & 0 & 0 & 0 \\ 1 & 0 & 1 & 0 & 0 & 0 & 0 & 0 \\ 1 & 0 & 1 & 0 & 0 & 0 & 0 & 0 \end{pmatrix}$$

are all symmetrically singular (in E'' we have omitted the ideal empty central row). For matrices with an odd number of rows, condition $p_e = p_o$ is violated; for matrices with even number m of rows (if $m \geq 4$), this singularity follows from Theorems A and α, since the matrix $\overline{E_1''}$ has a single odd supported sequence in its second row. Similar statements hold for many other Túran matrices.

3. ARBITRARY MATRICES

There exist similar theorems for non-symmetric matrices. In a matrix E, let i_o be a distinguished row, let i_o+s_j and i_o-s_j', $j = 1,\dots,r$, $0 < s_1 < \dots < s_r$, $0 < s_1' < \dots < s_r'$ be some other coupled rows. Let q_j, $j = 1,\dots,r$, denote the number of k's for which $e_{i_o-s_j',k} = e_{i_o+s_j,k} = 1$, in other words let q_j be the number of ones in identical positions in the two coupled rows. Let q_e, p_e (or q_o, p_o) be the numbers of ones and of zeros in even (or odd) positions in row i_o.

THEOREM B. *In the above assumptions, necessary for the regularity of* E *are the conditions*

$$q_1 + \dots + q_r \le p_e \quad , \tag{3.1}$$

$$q_1 + \dots + q_r \le p_o \quad . \tag{3.2}$$

Proof. We prove (3.1). If the matrix E is regular, then equations (1.1) have a solution for arbitrary c_{ik} and an arbitrary set of knots X . We select the x_i in such a way that

$$x_{i_o} = 0 \, , \quad y_j = x_{i_o + s_j} \, , \quad j = 1, \dots, r \, , \quad 0 < y_1 < \dots < y_r \, .$$

Then equations (1.1) give for the even part P_e of P (see also (2.1),(2.2))

$$\begin{aligned} P_e^{(k)}(y_j) &= \tfrac{1}{2}[c_{i_o+s_j,k} + (-1)^k c_{i_o-s_j',k}] \, , \quad e_{i_o+s_j,k} = 1 \, , \\ P_e^{(k)}(0) &= c_{i_o k} \, , \quad k \text{ even} \, , \qquad e_{i_o k} = 1 \quad . \end{aligned} \tag{3.3}$$

We have a system of $q_1 + \dots + q_j + q_e$ equations for the coefficients of P_e . The number of these coefficients is $[\frac{n}{2}] + 1$ - number of even positions in each row of n+1 elements. The inequality

$$q_e + q_1 + \dots + q_r \le [\tfrac{n}{2}] + 1$$

must be satisfied, for there must be solutions for arbitrary right hand sides in (3.3). But since $[\frac{n}{2}] + 1 - q_e = p_e$, we have (3.1).

REFERENCES

1. Ferguson, D. (1969). The Question of Uniqueness for G.D. Birkhoff Interpolation Problems, *J. Approx. Theory* 2, 1-28.

2. Lorentz, G.G. (1972). Birkhoff Interpolation and the Problem of Free Matrices, *J. Approx. Theory* 6, 283-290.

3. Lorentz, G.G. (1974). The Birkhoff Interpolation Problem: New Methods and Results, pp.481-501, *in*: Proceedings Int. Conference Oberwolfach, Birkhäuser Verlag, Basel (ISNM25).

4. Lorentz, G.G. (1977). Coalescence of Matrices, Regularity and Singularity of Birkhoff Interpolation Problems, *J. Approx. Theory* 20, 178-190.

5. Nemeth, A.B. (1966). Transformations of the Chebyshev Systems, *Math. Cluj* 8(31), 315-333.

6. Windhauer, H. (1971). Zur symmetrischen Lücken Interpolation, *Z. Angew. Math. Mech.* 51, T31-T32.

ACKNOWLEDGEMENT

This work has been supported, in part, by the grant MCS77-04946 of the National Science Foundation.

NEAR-BEST L_∞ AND L_1 APPROXIMATIONS TO ANALYTIC FUNCTIONS ON TWO-DIMENSIONAL REGIONS

J.C. Mason

Department of Mathematics and Ballistics
Royal Military College of Science
Shrivenham, Swindon, Wilts

ABSTRACT

For analytic functions of a complex variable, it is a well-known consequence of the maximum modulus theorem that approximation problems in the L_∞ norm on two-dimensional regions reduce to corresponding problems on one-dimensional contours. It also happens to be true that certain results in the L_p norms $(1 \le p < \infty)$ on two-dimensional regions may be derived from corresponding results for one-dimensional contours by integrating with respect to an appropriate variable. These two ideas are illustrated to prove a number of new results on near-best L_∞ and L_1 approximation on regions in the complex plane.

In the case of L_∞ approximation, it is proved that on a circular annulus the projection of a function on the partial sum of its Laurent series is a minimal projection on polynomials in z and z^{-1}, and that the resulting approximation is near-minimax. By applying a conformal mapping it is simple to deduce various known results concerning the uniform convergence and near-minimax approximation of Chebyshev series of the first kind of elliptical regions. In the case of L_1 approximation, earlier results obtained by the author for contours are integrated over a radial variable to yield new results for the regions interior to those contours. Specifically, it is proved that the partial sums of the Taylor series, Laurent series, and Chebyshev series of the second kind are near-best L_1 approximations with respect to weight functions 1, 1, and $|z^2-1|^{-\frac{1}{2}}$ respectively, on the regions bounded by a circle, an annulus, and an ellipse, respectively. The L_1 convergence of these three series is also established on the relevant regions.

1. INTRODUCTION

It has already been proved for analytic functions on the unit disc[1] that the projection on the partial sum of the Taylor series is minimal in L_∞ and that the resulting approximation is near-minimax. The corresponding results for the Laurent series on a circular annulus will now be established in a similar way and the transformation $z = w + \sqrt{(w^2-1)}$ applied to deduce the known result[2] that the partial sum of the Chebyshev series of the first kind is near-minimax on an elliptical region. Two other standard results will also be deduced: that the Chebyshev series converges uniformly on the interior of the ellipse and that it coincides with the Chebyshev series for the real interval $[-1,1]$ when the complex variable z is replaced by the real variable x.

In recent work[3] on the L_1 norm it was proved that the Taylor series partial sum is near-best in L_1 on the circular contour $|z| = \rho$, and the analogous result for the Laurent series partial sum will now be proved similarly. It was also shown in [3] that the partial sum of the Chebyshev series of the second kind is near-best and converges in L_1 on the elliptical contour $|w + \sqrt{(w^2-1)}| = \rho$. By integrating these three results over ρ we are able to deduce new results on near-best L_1 approximation and L_1 convergence for the regions interior to a circle, an annulus, and an ellipse.

For completeness, the results on near-best L_∞ and L_1 approximation will be compared with known results on best L_2 approximation for the same contours and regions.

2. NEAR-BEST APPROXIMATION

Suppose that $R = \overline{I(\Gamma)}$ is a closed region with boundary Γ and interior $I(\Gamma)$, and that $A(\Gamma)$ is the linear space of functions $f(z)$ analytic in $I(\Gamma)$ and continuous on R normed by some chosen norm on R.

Following [1], a polynomial approximation f_n of degree n to f on R is said to be near-best within a relative distance γ if

$$\|f - f_n\| \leq (1 + \gamma) \cdot \|f - f_n^B\|$$

where f_n^B is a best approximation of degree n. Here f_n is only a practical near-best approximation if γ is acceptably small, and an asymptotic behaviour of $\log n$ for γ is typically satisfactory.

A projection P is a bounded, linear, idempotent operator, and always has the property (see [1]) that

$$\|f - Pf\| \leq (1 + \|P\|) \cdot \|f - f_n^B\| \quad .$$

Thus Pf is a near-best approximation within a relative distance $\|P\|$, which will have practical meaning if $\|P\|$ is small. A projection for which $\|P\|$ is smallest is termed a minimal projection.

A near-best approximation in the L_∞ norm is called a near-minimax approximation.

3. L_∞ APPROXIMATION BY TAYLOR AND LAURENT SERIES

If we define

$$\|f\| = \|f\|_\infty = \max_R |f| \quad ,$$

then we know by the maximum modulus theorem that

$$\|f\|_\infty = \max_\Gamma |f| \quad .$$

Thus we need only measure norms over the boundary Γ .

Let Π_n be the space of polynomials of degree n in z , and choose Γ to be the circle C_ρ: $|z| = \rho$. Denote by S_n the projection of a function in $A(C_\rho)$ onto the partial sum of degree n of its Taylor series expansion. The following two results were established in [1], and the third result is standard.

THEOREM 3.1 *S_n is a minimal projection of $A(C_\rho)$ onto Π_n.*

THEOREM 3.2 *$S_n f$ is a practical near-minimax approximation on C_ρ.* Specifically

$$\|S_n\|_\infty \leq \tau_n \quad ,$$

where

$$\tau_n = \frac{1}{\pi}\int_0^\pi \frac{|\sin(n+1)x|}{\sin x}\,dx \tag{1}$$

and

$$\tau_n \sim \frac{4}{\pi^2}\log n \quad . \tag{2}$$

THEOREM 3.3 *$S_n f$ converges uniformly to f on any closed region interior to C_ρ.*

Now let Π_n^* be the space of polynomials of degree n in z and z^{-1}, and choose Γ to be the union C_ρ^* of the two concentric circles $|z| = \rho > 1$ and $|z| = \rho^{-1}$. (Any pair of concentric circles can be so chosen by a suitable transformation.) Let $I(C_\rho^*)$ be the annulus N_ρ: $\{\rho^{-1} < |z| < \rho\}$, let $A(C_\rho^*)$ denote the class of functions analytic in N_ρ and continuous on C_ρ^*, and denote by B_n the projection of a function in $A(C_\rho^*)$ onto the partial sum of order n of its Laurent series expansion. Then the following three theorems hold for B_n on $A(C_\rho^*)$, and are analogous to Theorems 3.1, 3.2, and 3.3. The proofs of Theorems 3.4 and 3.5 are very similar to those of 3.1 and 3.2 (in [1]), and rely on the fact that the unit circle is interior to $I(\Gamma)$. Theorem 3.6 is standard.

THEOREM 3.4 *B_n is a minimal projection of $A(C_\rho^*)$ onto Π_n^*.*

Proof of 3.4 Let P_n be any projection of $A(C_\rho^*)$ onto Π_n^*. Then we must prove that

$$||B_n||_\infty \leq ||P_n||$$

For any fixed t on C_1: $|z| = 1$, define the shift operator E_t: $A(C_\rho^*) \rightarrow A(C_\rho^*)$ by

$$(E_t f)(z) = f(tz) \quad ,$$

and define an operator ψ by

$$(\psi f)(z) = \frac{1}{2\pi i} \int_{C_1} (E_{\bar{t}} P_n E_t f)(z)\ t^{-1}\, dt \quad , \tag{3}$$

where $\bar{t} = t^{-1}$. We shall prove that $\psi = B_n$, and to do this it will be sufficient to show that for integral k

$$\psi f_k = B_n f_k \quad , \qquad -\infty < k < \infty$$

where

$$f_k = z^k \quad ,$$

since the positive and negative powers of z form a basis in $A(C^*_\rho)$.

(i) If $-n \le k \le n$, $f_k \in \Pi^*_n$ and so $B_n f_k = f_k$.
Also $E_t f_k = t^k z^k \in \Pi^*_n$, so that $P_n E_t f_k = E_t f_k$.
Hence $E_{\bar{t}} P_n E_t f_k = f_k$, since $E_{\bar{t}} E_t = I$.
Thus

$$(\psi f_k)(z) = \frac{1}{2\pi i} \int_{C_1} (E_{\bar{t}} P_n E_t f_k)(z) \frac{dt}{t} = f_k(z)$$

and

$$\psi f_k = B_n f_k \quad \text{for} \quad -n \le k \le n \quad .$$

(ii) If $k > n$ or $k < -n$, $B_n f_k = 0$.
Also $(E_{\bar{t}} P_n E_t f_k)(z) = t^k (E_{\bar{t}} P_n f_k)(z) = t^k (P_n f_k)(t^{-1} z)$,
and $(P_n f_k)(w) = \sum_{j=-n}^{n} c_j w^j$ for some coefficients c_j
since $P_n f_k \in \Pi^*_n$.
For $k > n$, it follows from (3) that

$$(\psi f_k)(z) = \frac{1}{2\pi i} \int_{C_1} t^{k-n-1} \{ \sum_{j=-n}^{n} c_j t^{n-j} z^j \}\, dt \quad ,$$

and the integral is zero since the integrand is a polynomial in t .

For $k < -n$, it follows from (3) that

$$(\psi f_k)(z) = \frac{1}{2\pi i}\int_{C_1} t^{k+n-1}\{\sum_{j=-n}^{n} c_j t^{-j-n} z^j\}\, dt \quad ,$$

and the integral is zero since the integrand is a polynomial in t^{-1} in which the lowest power of t^{-1} occurring is at least 2 .

Hence $\psi f_k = B_n f_k$ (= 0) for $k > n$ or $k < -n$.

It thus follows that $\psi = B_n$ identically. But, from the definition of ψ ,

$$||\psi f||_\infty \le ||E_{\bar{t}} P_n E_t f||_\infty \cdot \frac{1}{2\pi}\int_{C_1} |dt| \le ||P_n||_\infty \cdot ||f||_\infty$$

since E_t has norm 1 .

Hence

$$||B_n||_\infty = ||\psi||_\infty \le ||P_n||_\infty$$

Q.E.D.

THEOREM 3.5 *$B_n f$ is a practical near minimax approximation to f on C^*_ρ .*

Specifically

$$||B_n||_\infty \le \lambda_n \quad ,$$

where

$$\lambda_n = \frac{1}{\pi}\int_0^{\pi} \frac{|\sin(n+\frac{1}{2})x|}{\sin \frac{1}{2}x}\, dx \tag{4}$$

and

$$\lambda_n \sim \frac{4}{\pi^2} \log n \quad . \tag{5}$$

Proof of 3.5 For any fixed z in N_ρ, the partial sum of the Laurent series may be expressed (by using Cauchy's integral formulae for the coefficients, taken around C_r: $|t| = |z|$) in the form

$$(B_n f)(z) = \frac{1}{2\pi i} \int_{C_r} \sum_{k=-n}^{n} \left(\frac{z}{t}\right)^k \frac{f(t)}{t} \, dt \quad .$$

Transforming to the new variable $s = tz^{-1}$, and using the relation

$$\sum_{k=-n}^{n} s^{-k} = s^{-n} \frac{s^{2n+1} - 1}{s - 1} \quad ,$$

we deduce that

$$(B_n f)(z) = \frac{1}{2\pi i} \int_{C_1} \frac{s^{2n+1} - 1}{s^{n+1}(s-1)} f(zs) \, ds \quad . \tag{6}$$

Thus

$$||B_n f||_\infty \le \frac{1}{2\pi} \int_{C_1} \frac{|s^{2n+1} - 1|}{|s - 1|} |ds| \cdot ||f||_\infty \quad .$$

Now

$$\frac{1}{2\pi} \int_{C_1} \frac{|s^{2n+1} - 1|}{|s-1|} |ds| = \frac{1}{\pi} \int_0 \frac{|\sin(n+\frac{1}{2})x|}{\sin \frac{1}{2}x} dx \quad ,$$

where $s = e^{ix}$ (compare [1]) and hence

$$||B_n f||_\infty \le \lambda_n ||f||_\infty \quad .$$

Thus $\|B_n\|_\infty \le \lambda_n$, and, since λ_n is the nth Lebesgue constant, the asymptotic behaviour (5) is well known (compare [2]).

Q.E.D.

THEOREM 3.6 *$B_n f$ converges uniformly to f on any closed region interior to C_ρ.*

4. L_∞ APPROXIMATION BY CHEBYSHEV SERIES

Suppose $f(w)$ is in the space $A(\xi_\rho)$ of functions analytic on the interior $I(\xi_\rho)$ of the ellipse

$$\xi_\rho : |w + \sqrt{(w^2 - 1)}| = \rho > 1$$

and continuous on ξ_ρ. Then the function

$$g(z) = 2f(\tfrac{1}{2}(z + z^{-1})) \tag{7}$$

is analytic on the annulus

$$N_\rho : \rho^{-1} < |z| < \rho$$

and continuous on its boundary C^*_ρ, and the transformation

$$w = \tfrac{1}{2}(z + z^{-1}) \tag{8}$$

takes $g(z)$ into $2f(w)$. Hence $G(z)$ has a Laurent series expansion on N_ρ, with a partial sum of the form

$$(B_n g)(z) = \sum_{k=-n}^{n} a_k z^k .$$

It is clear from (7) that a_k and a_{-k} are equal, and hence by Cauchy's integral formula

$$(B_n g)(z) = \sum_{k=0}^{n}{}' a_k (z^k + z^{-k}) ,$$

where

$$a_k = \frac{1}{4\pi i} \int_{C_r} g(z)(z^k + z^{-k}) \frac{dz}{z} .$$

The dash denotes that the first term is halved, and the integral may be taken over C_r: $|z| = r$ for any r such that $1 < r \leq \rho$.

Now for $1 < |z| \leq \rho$, (8) has the inverse mapping

$$z = w + \sqrt{(w^2 - 1)} .$$

Also $\frac{1}{2}(z^k + z^{-k}) = T_k(w)$, the Chebyshev polynomial of the first kind of degree k in w , and

$$\frac{dz}{z} = \frac{dw}{\sqrt{(w^2 - 1)}} .$$

Hence

$$(B_n g)(z) = \sum_{k=0}^{n}{}' a_k (z^k + z^{-k}) = 2 \sum_{k=0}^{n}{}' a_k T_k(w) ,$$

where

$$g(z) = 2f(w)$$

and

$$a_k = \frac{1}{4\pi i} \int_{C_r} g(z)(z^k + z^{-k}) \frac{dz}{z}$$

$$= \frac{1}{\pi i} \int_{\xi_r} f(w) T_k(w)(w^2 - 1)^{-\frac{1}{2}} dw \qquad \text{for } 1 < r \leq \rho \qquad (9)$$

We note (compare [1]) that (9) is precisely the formula for the coefficient of $T_k(w)$ in the Chebyshev series expansion of $f(w)$ on $I(\xi_r)$.

Thus the Laurent series of $g(z)$ on N_ρ is precisely twice the Chebyshev series of $2f(w)$ on $I(\xi_\rho)$ for any $\rho > 1$. Moreover, if G_n denotes the projection of a function on the partial sum of its Chebyshev series expansion of the first kind on $I(\xi_\rho)$, then

$$(B_n g)(z) = 2(G_n f)(w) \quad . \tag{10}$$

We may now deduce three known results for Chebyshev series on ellipses. The constant λ_n , defined in (4) above, features once more.

THEOREM 4.1 *$G_n f$ is a practical near-minimax approximation to f on ξ_ρ. Specifically* $\|G_n\|_\infty \le \lambda_n$.

Proof of 4.1 By Theorem 3.5

$$\|(B_n g)(z)\|_\infty \le \lambda_n \|g(z)\|_\infty \quad ,$$

and from (7), (8), and (10) it follows that

$$2\,\|(G_n f)(w)\|_\infty \le 2\lambda_n \|f(w)\|_\infty \quad .$$

Hence

$$\|G_n f\|_\infty \le \lambda_n \|f\|_\infty \quad .$$

Q.E.D.

THEOREM 4.2 *$G_n f$ converges uniformly to f on any closed region interior to ξ_ρ.*

Proof of 4.2 The Laurent series of g converges uniformly on $r^{-1} \le |z| \le r$ for any $r < \rho$, by Theorem 3.6. By the relations (7), (8), and (10) it follows that the Chebyshev series of f converges uniformly on $|w + \sqrt{(w^2-1)}| \le r$.

Q.E.D.

THEOREM 4.3 *The Chebyshev series expansion on $I(\xi_r)$ is identical for every r $(1 < r \le \rho)$, and when the variable w is replaced by the real variable x it coincides with the Chebyshev series expansion on [-1,1].*

Proof of 4.3 Since the coefficient of $T_k(w)$ in the Chebyshev series expansion of f on $I(\xi_\rho)$ is given by (9) for every r $(1 < r \le \rho)$, it follows that the expansion on $I(\xi_r)$ is identical for every r. Now in the limit as $r \to 1$, the contour ξ_r becomes the real interval [-1,1] taken in each direction, while the square root $(w^2 - 1)^{\frac{1}{2}}$ becomes $\pm i \sqrt{(1-x^2)}$ in the respective directions. Hence

$$
\begin{aligned}
a_k &= \lim_{r\to 1} \frac{1}{\pi i} \int_{\xi_r} f(w) T_k(w) (w^2-1)^{-\frac{1}{2}} \, dw \\
&= \frac{2}{\pi} \int_{-1}^{1} f(x) T_k(x) (1-x^2)^{-\frac{1}{2}} \, dx ,
\end{aligned}
$$

which is precisely the coefficient of $T_k(x)$ in the Chebyshev series expansion of f(x) on [-1,1].

Q.E.D.

We note that Theorem 4.1 is a result previously obtained by Geddes[2]. In his approach, the theory of Faber polynomials was used under an exterior transformation of C_ρ to deduce the expression (9) for the Chebyshev series coefficients. The theorem was then obtained by summing the Chebyshev series and using various trigonometric relationships to bound the norm of the projection.

5. L_1 APPROXIMATIONS ON CONTOURS

Consider now functions $f(z)$ in $A(\Gamma)$ measured in the L_p norm on Γ. Then for $1 \le p < \infty$,

$$\|f(z)\| = \|f(z)\|_p^\Gamma = \{\int_\Gamma |f(z)|^p \, |dz|\}^{1/p} \quad . \tag{11}$$

Denote again by τ_n and λ_n the constants (1) and (4) encountered in Theorems 3.2, 3.5.

A pair of results on near-best L_1 approximation and L_1 convergence, respectively, were established in [3] for the Taylor series projection S_n on the circle C_ρ: $|z| = \rho > 1$. These may be trivially generalised to the interior circle C_r: $|z| = r$ for any r $(0 < r \le \rho)$ to give the following theorem.

THEOREM 5.1 *If f is in $A(C_\rho)$, then for $0 < r \le \rho$*

(a) $S_n f$ is a practical near-best L_1 approximation to f on C_r. Specifically $\|S_n\|_1^{C_r} \le \tau_n$.

(b) $S_n f$ converges in L_1 to f on C_r.

A corresponding pair of results for the Laurent series projection B_n may be proved similarly on C_r interior to C_ρ^*: $|z| = \rho$, ρ^{-1}.

THEOREM 5.2 *If f is in $A(C_\rho^*)$, then for $\rho^{-1} \le r \le \rho$*

(a) $B_n f$ is a practical near-best L_1 approximation to f on C_r. Specifically

$$\|B_n\|_1^{C_r} \le \lambda_n \quad .$$

(b) $B_n f$ converges in L_1 to f on C_r.

Proof of 5.2

(a) From (6) above,

$$(B_n f)(z) = \frac{1}{2\pi i}\int_{C_1} \frac{s^{2n+1}-1}{s^{n+1}(s-1)}\, f(zs)\, ds$$

Hence

$$\|B_n f\|_1^{C_r} = \frac{1}{2\pi}\int_{C_r} \{\left|\int_{C_1} \frac{s^{2n+1}-1}{s^{n+1}(s-1)}\, f(zs)\, ds\right|\}\, |dz|$$

$$\le \frac{1}{2\pi}\int_{C_r} \{\int_{C_1} \frac{|s^{2n+1}-1|}{|s-1|}\, |f(zs)|\, |ds|\}\, |dz|$$

$$\le \frac{1}{2\pi}\int_{C_1} \frac{|s^{2n+1}-1|}{|s-1|}\, |ds| \cdot \int_{C_r} |f(t)|\, |dt|$$

(reversing the order of integration and setting $t = zs$).
i.e.

$$\|B_n f\|_1^{C_r} \le \lambda_n \|f\|_1^{C_r} \qquad \text{from (4).}$$

The result follows.

(b) L_1 convergence follows immediately from L_2 convergence, which is a consequence of the orthogonality of the powers of z on C_r.

Q.E.D.

Finally a corresponding pair of results in [3] for the projection H_n of a function on the partial sum of degree n of its expansion in Chebyshev polynomials $\{U_k(w)\}$ of the second kind of the interior $I(\xi_\rho)$ of the ellipse ξ_ρ: $|w + \sqrt{(w^2-1)}| = \rho$ may be trivially generalised to the interior ellipse ξ_r: $|w + \sqrt{(w^2-1)}| = r$ $(1 < r \le \rho)$ to give the following theorem.

THEOREM 5.3 *If f is in $A(\xi_\rho)$, then for $1 < r \le \rho$*

(a) $H_n f$ is a practical near-best L_1 approximation to f on ξ_r. Specifically

$$||H_n||_1^{\xi_r} \le \lambda_{n+1} \quad .$$

(b) $H_n f$ converges in L_1 to f on ξ_r.

6. L_1 APPROXIMATIONS ON REGIONS

Suppose that for f in $A(\Gamma)$ we adopt an L_p norm with weight function $w(z)$ on the two-dimensional region $R = R(\Gamma) = \overline{I(\Gamma)}$. Then, for $1 \le p < \infty$,

$$||f(z)|| = ||f(z)||_p^R = \{\int_{R(\Gamma)} w(z) \; |f(z)|^p \; dS\}^{1/p} \tag{12}$$

where dS is an element of area.

For circular, annular and elliptical regions we may relate this norm to the norm (11) on a class of interior contours as follows. (The contours C_r, C_ρ, C^*_ρ, ξ_r, ξ_ρ are as defined in §5.)

LEMMA 6.1 *If $w(z) = 1$, $\Gamma = C_\rho$ and $z = re^{i\theta}$, then*

$$dS = |dz| \cdot dr$$

and

$$\{\|f(z)\|_p^R\}^p = \int_0^\rho \{\|f(z)\|_p^{C_r}\}^p \, dr$$

LEMMA 6.2 *If $w(z) = 1$, $\Gamma = C_\rho^*$ and $z = re^{i\theta}$, then*

$$dS = |dz| \cdot dr$$

and

$$\{\|f(z)\|_p^R\}^p = \int_{\rho^{-1}}^\rho \{\|f(z)\|_p^{C_r}\}^p \, dr$$

LEMMA 6.3 *If $w(z) = |z^2-1|^{-\frac{1}{2}}$, $\Gamma = \xi_\rho$,*

and

$$z = \tfrac{1}{2}(r + r^{-1}) \cos\theta + i\tfrac{1}{2}(r - r^{-1}) \sin\theta \qquad (13)$$

then

$$dS = |z^2 - 1|^{\frac{1}{2}} \, r^{-1} \, |dz| \, dr$$

and

$$\{\|f(z)\|_p^R\}^p = \int_1^\rho \{\|f(z)\|_p^{\xi_r}\}^p \frac{dr}{r} \quad .$$

Proof of Lemmas The first two lemmas follow immediately from the definitions of norms and the relation

$$dS = r \, dr \, d\theta \quad .$$

For Lemma 6.3, the parametrisation (13) of ξ_r gives

$$dS = \tfrac{1}{4}r^{-1}(r^2 + r^{-2} - 2\cos 2\theta) \, dr \, d\theta \quad .$$

But

$$|z^2 - 1| = \tfrac{1}{4}(r^2 + r^{-2} - 2 \cos 2\theta) \quad ,$$

and for any fixed r

$$dz = \tfrac{1}{2}(r^2 + r^{-2} - 2 \cos 2\theta)^{\frac{1}{2}} \, d\theta \quad .$$

Hence the result follows from the definitions and the choice of w .

Q.E.D.

We now derive results analogous to Theorems 5.1, 5.2, and 5.3 for the three projections S_n, B_n, and H_n .

THEOREM 6.4 *If f is in $A(C_\rho)$, $R = R(C_\rho)$, and $w(z) = 1$, then*

(a) $S_n f$ is a practical near-best L_1 approximation to f on R . Specifically

$$||S_n||_1^R \leq \tau_n \quad .$$

(b) $S_n f$ converges in L_1 to f on R .

Proof of 6.4 From Theorem 5.1

$$||S_n f||_1^{C_r} \leq \tau_n ||f||_1^{C_r}$$

Integrating over r from 0 to ρ and applying Lemma 6.1 for $p = 1$, we deduce that

$$||S_n f||_1^R \leq \tau_n ||f||_1^R \quad .$$

This proves (a), and (b) follows similarly.

THEOREM 6.5 *If f is in $A(C_\rho^*)$, $R = R(C_\rho^*)$, and $w(z) = 1$, then*

(a) $B_n f$ is a practical near-best L_1 approximation to f on R. Specifically

$$\|B_n\|_1^R \le \lambda_n \quad .$$

(b) $B_n f$ converges in L_1 to f on R.

Proof of 6.5 From Theorem 5.2

$$\|B_n f\|_1^{C_r} \le \lambda_n \|f\|_1^{C_r} \quad .$$

Integrating over r from ρ^{-1} to ρ and applying Lemma 6.2 for $p = 1$, we deduce that

$$\|B_n f\|_1^R \le \lambda_n \|f\|_1^R \quad .$$

This proves (a), and (b) follows similarly.

THEOREM 6.6 *If f is in $A(\xi_\rho)$, $R = R(\xi_\rho)$, and $w(z) = |z^2-1|^{-\frac{1}{2}}$, then*

(a) $H_n f$ is a practical near-best L_1 approximation to f on R. Specifically

$$\|H_n\|_1^R \le \lambda_{n+1} \quad .$$

(b) $H_n f$ converges in L_1 to f on R.

Proof of 6.6 From Theorem 5.3

$$\|H_n f\|_1^{\xi_r} \leq \lambda_{n+1} \|f\|_1^{\xi_r}$$

Multiplying through by r^{-1}, integrating over r from 1 to ρ, and applying Lemma 6.3 for $p = 1$, we deduce that

$$\|H_n f\|_1^R \leq \lambda_{n+1} \|f\|_1^R \ .$$

This proves (a), and (b) follows similarly.

7. A COMPARISON WITH RESULTS IN L_2

The partial sums of order n of the Taylor series, Laurent series, and Chebyshev series of first and second kinds are in fact all best L_2 approximations with respect to appropriate weight functions on all the contours and regions considered above. Most of the relevant results are given in Davis[4], and the remainder may easily be derived. We note in particular that Lemmas 6.1, 6.2, and 6.3 immediately yield results for regions from results for contours.

A summary of all the best L_2 approximations and appropriate weight functions is given in Table 1, together with all the corresponding near-best L_∞ and L_1 approximations and appropriate weight functions. Some of the L_∞ and L_1 results tabulated (for Ch_2 in L_∞ and Ch_1 in L_1) are included for completeness, although they have not been discussed above.

Ideally we should like a norm with a unit weight function. This is achieved by the Taylor and Laurent series on circular and annular contours and regions in L_∞, L_1 and L_2. It is also achieved by the Chebyshev series of the first kind on an elliptical contour and elliptical region in L_∞, and by the Chebyshev series of the second kind on an elliptical contour in L_1 and on an elliptical region in L_2. It is not achieved by any of the series on an elliptical region in L_1 or on an elliptical contour in L_2.

Table I Series Yielding Best and Near-best Approximations for Appropriate Weight Functions w(z)

Domain	L_∞ (Near-best)		L_1 (Near-best)		L_2 (Best)	
	Series	w(z)	Series	w(z)	Series	w(z)
Circle						
a) Contour	T	1	T	1	T	1
b) Region	T	1	T	1	T	1
Annulus						
a) Contour	L	1	L	1	L	1
b) Region	L	1	L	1	L	1
Ellipse						
a) Contour	Ch_1	1	Ch_1	$\lvert z^2-1\rvert^{-\frac{1}{2}}$	Ch_1	$\lvert z^2-1\rvert^{-\frac{1}{2}}$
"	Ch_2	$\lvert z^2-1\rvert^{\frac{1}{2}}$	Ch_2	1	Ch_2	$\lvert z^2-1\rvert^{\frac{1}{2}}$
b) Region	Ch_1	1	Ch_1	$\lvert z^2-1\rvert^{-1}$	Ch_1	$\lvert z^2-1\rvert^{-1}$
"	Ch_2	$\lvert z^2-1\rvert^{\frac{1}{2}}$	Ch_2	$\lvert z^2-1\rvert^{-\frac{1}{2}}$	Ch_2	1

T = Taylor series, L = Laurent series,
Ch_1, Ch_2 = Chebyshev series of 1st, 2nd kinds.

8. APPENDIX - TAYLOR PROJECTION CONSTANT

We are indebted to Dr T.J. Rivlin for pointing out to us that the norm of the Taylor projection, $\|S_n\|_\infty$, can be estimated more precisely than we have done in Theorem 3.2 above. A result of Landau (1913-1916), which is discussed in detail in [5], establishes that

$$\|S_n\|_\infty = \sum_{k=0}^{n} \binom{-\frac{1}{2}}{k}^2 \sim \frac{1}{\pi} \log n \quad .$$

REFERENCES

1. Geddes, K.O. and Mason, J.C. (1975). Polynomial approximation by projections on the unit circle, *SIAM J. Numer. Anal.* 12, 111-120.

2. Geddes, K.O. (1977). Near-minimax polynomial approximations in an elliptical region, *SIAM J. Numer. Anal.* (In press.)

3. Mason, J.C. (1977). Near-best L_1 approximations on circular and elliptical contours. (Submitted to *J. of Approx. Theory.*)

4. Davis, P.J. (1963). Interpolation and Approximation, Blaisdell.

5. Dienes, P. (1931). The Taylor Series - An Introduction to the Theory of Functions of a Complex Variable, Oxford.

INTERPOLATIVE SUBSPACES, INTERPOLATIVE WIDTHS AND INTERPOLATIVE MAPS

Geetha S. Rao

Ramanujan Institute, University of Madras
Madras, India

1. INTRODUCTION

It is well known that approximation of a given class of functions by algebraic or trigonometric polynomials is a very special process. Abstract approximation seeks to find other favourable systems of approximants. With this end in view, the concept of the deviation of one set from another has been defined.

Let X be a Banach space and let A,B be subsets of X.

Definition The *deviation of* A *from* B is the number

$$\delta(A,B) = \sup_{x\in A} \inf_{y\in B} \|x-y\| \quad .$$

Kolmogorov[4] considers the case when $B = X_n$, an n-dimensional subspace of X and examines the possibility of estimating the infimum of this deviation for all n-dimensional subspaces X_n of X, if it exists.

Definition The *n-width of* A *in* X is given by

$$d_n^X(A) \quad d_n(A) = \inf_{X_n} \delta(A,X_n) \quad , \quad n=0,1,2,\ldots \quad .$$

Evidently, $d_n(A) = 0$ for each A, if $\dim X = n$. If $\dim X < n$, $d_n(A)$ is defined to be zero.

Let Φ be a linear subspace of X^*, the dual of X.

Definition A linear subspace $L \subset X$ is Φ-*interpolative* if for every $x \in X$ there exists one and only one $y \in L$ such that $\phi(x) = \phi(y)$ for all $\phi \in \Phi$.

The mapping $J_L : x \to y$ is called a Φ-*interpolative map*. Obviously J_L is a projection onto L.

To relate the notion of Φ-interpolative subspace with the notion of n-width, the concept of Φ-interpolative width is introduced[6].

Definition Let L be a Φ-interpolative subspace of X and let $M \subset X$. Define

$$\sigma_\Phi(M,L) = \sup_{x \in M} \|x - J_L x\|$$

and

$$\sigma_\Phi(M) = \inf_L \sigma_\Phi(M,L) \quad .$$

Then $\sigma_\Phi(M)$ is the Φ-*interpolative width* of M.

2. Φ-INTERPOLATIVE SUBSPACES

It is interesting to investigate the existence of such subspaces and to see what happens if they are closed. The following details are due to Milota[6].

The first theorem provides a characterization of these subspaces.

THEOREM L is Φ-interpolative if and only if $X = L \oplus \Phi_\perp$.

The next theorem indicates a criterion for L to be closed.

THEOREM Let L be a Φ-interpolative subspace of X. Then L is closed if and only if $X^* = (\Phi_\perp)^\perp \oplus L^\perp$.

Observe that if L is closed, $X^* = (\Phi_\perp)^\perp \oplus L^\perp$ is a necessary condition for L to be Φ-interpolative. If X is *reflexive*, then this condition is sufficient as well.

Denote by J_L^* the adjoint corresponding to J_L.

THEOREM If L is a closed Φ-interpolative subspace of X, then J_L^* is the projection onto $(\Phi_\perp)^\perp$ which is parallel to $L^\perp$.

No restraint was placed on the dimension of Φ. It turns out that if Φ is finite dimensional, certain characterizations are possible.

THEOREM If Φ is finite dimensional, the following are equivalent:

a) L is Φ-interpolative

b) there exists a basis $\{x_i\}_{i=1}^n$ of L such that

$$\phi_j(x_i) = \delta_{ij}, \quad i,j = 1,2,\ldots,n$$

c) $x = L + \Phi_\perp$ and $\dim L = n$

d) $X^* = \Phi \oplus L^\perp$.

The notion of best Φ-interpolative subspace is now defined.

Definition If there exists a closed Φ-interpolative subspace $\tilde{L}$ of X such that

$$\|J_{\tilde{L}}\| = \inf_{L} \|J_L\| \quad ,$$

then $\tilde{L}$ is called the *best* Φ-*interpolative subspace*.

The existence and characterization of such best interpolative subspaces, analogous to the results of Aubin[1], are established below. The finite dimensionality of Φ and the reflexive nature of X are crucial, for in the case of an arbitrary Banach space or a subspace Φ of arbitrary dimension, nothing is known about the existence.

THEOREM Let X be a reflexive Banach space and let Φ be a finite dimensional subspace of X^* . Then there exists the best Φ-interpolative subspace.

For the characterization, while the dimension of Φ is not required to be finite, another condition is placed on Φ .

THEOREM Let X be a reflexive Banach space and let Φ be a subspace of X^* such that $(\Phi_\perp)^\perp$ admits a bounded projection onto itself. Then $\tilde{L}$ is the best Φ-interpolative subspace if and only if J_L^* is a projection onto $(\Phi_\perp)^\perp$ with the smallest possible norm.

It is worthwhile to analyse the case when X is strictly convex, for, with the finite dimensionality of Φ , it can be conjectured that significant results will accrue (from a study of problems concerning SAIN by the author).

3. Φ-INTERPOLATIVE WIDTHS

If the dimension of Φ is finite,

$$d_n(M) \le \sigma_\Phi(M)$$

and if, in addition, M is bounded, then $\sigma_\Phi(M)$ is finite.

Using a technique similar to that of Garkavi[3], Milota[6] proves the existence of a Φ-interpolative subspace $\underset{\sim}{L}$ such that $\sigma_\Phi(M,\underset{\sim}{L}) = \sigma_\Phi(M)$ (= finite quantity), when $\dim \Phi < \infty$ and M is bounded.

In general, if L is a closed Φ-interpolative subspace of X, the deviation of M from L and $\sigma_\Phi(M,L)$ are related as follows:

$$\delta(M,L) \le \sigma_\Phi(M,L) \le (1 + \|J_L\| \delta(M,L)) \quad .$$

Other properties of Φ-interpolative widths can be found in [6].

4. Φ-INTERPOLATIVE MAPS AND RELATED IDEAS

Let π be the metric projection in X onto $\Phi_\perp$ and let m denote the collection of Φ-interpolative maps. The main theorem of Milota[7] is very instructive and is provided here.

Definition For $x \in X$, define

$$S(x) = \{y \in X : \exists\, \tilde{J} \in M \text{ with } \tilde{J}x = y \text{ and } \|y\| = \min_{J \in M} \|Jx\|\} \, .$$

and

$$S = \bigcup_{x \in X} S(x) \quad .$$

THEOREM

a) $y \in S(x)$ if and only if $y \in x - \pi x$

b) $S(x)$ is nonempty for all $x \in X$ if and only if $\Phi_\perp$ is proximinal and also if and only if $X = S + \Phi_\perp$

c) $S(x)$ contains no more than one element for all $x \in X$ if and only if $\Phi_\perp$ is semi-Chebyshev and also if and only if $(\mathcal{S}-\mathcal{S}) \cap \Phi = \{0\}$

d) $\mathcal{S}(x)$ contains exactly one element for all $x \in X$ if and only if $\Phi_\perp$ is a Chebyshev subspace and also if and only if for any $x \in X$ there exist $s \in \mathcal{S}$ and $z \in \Phi_\perp$, uniquely determined, such that $x = s + z$.

The continuity of J_L is discussed now.

THEOREM

a) If L is closed and Φ-interpolative, J_L is continuous

b) J_L is continuous if and only if $\inf\limits_{\substack{x \in L \\ \|x\| = 1}} \{\delta(x, \Phi_\perp)\} = \alpha > 0$.

In this case, $\|J_L\| = \frac{1}{\alpha}$.

Definition A Φ-interpolative map $\tilde{J}$ is *minimal* if the inequality

$$\|\tilde{J}x\| \leq \|Jx\|$$

holds for all $x \in X$ and $J \in m$.

The following gives a characterization of minimal Φ-interpolative maps.

THEOREM Let J be a Φ-interpolative map. The following are equivalent:

a) J is a minimal Φ-interpolative map

b) $\|J\| = 1$

c) $J(X) \subset \mathcal{S}$.

It must be observed that when X is reflexive and Φ is finite dimensional, then there exists a Φ-interpolative map with a minimal norm, as already seen in Section 2. If X is strictly convex, a minimal Φ-interpolative map exists if and only if the metric projection onto $\Phi_\perp$ is linear and this map is unique! For applications, refer to Milota[7].

5. CONCLUDING REMARKS

The theory described, using interpolative subspaces, is interesting and various questions related to approximation with interpolatory constraints may perhaps be resolved using it.

The choice of Φ is significant and more attention should be devoted to this aspect.

The techniques, though abstract in nature, seem worthwhile for applications.

REFERENCES

1. Aubin, J.P. (1968). Interpolation et Approximation Optimales et 'Spline Functions', *J. Math. Anal.* 24, 1-24.
2. Day, M.M. (1958). "Normed Linear Spaces", Springer.
3. Garkavi, A.L. (1962). On the Best Net and the Best Section of a Set in a Normed Linear Space (in Russian), *Izv. Akad. Nauk SSR*, ser. mat. 26, 87-106.
4. Kolmogorov, A.N. (1936). Über die beste Annäherung von Funktionen einer gegebenen Funktionenklasse, *Ann. of Math.* 37, 107-110.
5. Lorentz, G.G. (1966). "Approximation of Functions", Holt, Rinehart and Winston.
6. Milota, J. (1976). Interpolation in a Banach Space, *Czech. Math. J.* 26, 84-92.
7. Milota, J. (1977). Minimal Interpolation in a Normed Linear Space, *Indag. Math.* 39, 40-54.
8. Singer, I. (1970). "Best Approximation in Normed Linear Spaces by Elements of Linear Subspaces", Springer.
9. Tihomirov, V.M. (1960). Widths of Sets in Functional Spaces and the Theory of Best Approximations (in Russian), *Uspehi mat. nauk* 15, No.3, 81-120.

A GENERAL REMES ALGORITHM IN REAL OR COMPLEX NORMED LINEAR SPACES

H.-P. Blatt

Fakultät für Mathematik und Informatik
Universität Mannheim
Mannheim, Germany

1. INTRODUCTION

In 1934, Remes[5] developed an exchange algorithm for the approximation of continuous functions by polynomials in the Tchebycheff sense. Stiefel[7] described in 1959 an exchange method for the discrete Tchebycheff problem if the Haar condition is satisfied. For multivariate approximation Rice[6] has defined in 1963 the strict approximation to single out one of the best approximations as the "best of the bests". In 1965, Töpfer[8] presented a generalization of the Remes algorithm for the uniform approximation of continuous functions by elements of linear subspaces, if the Haar condition fails. This method is recursive with respect to the dimension of the linear approximating subspace. Its convergence was proved by Carasso and Laurent[2] in 1973. Last year Carasso and Laurent proposed a modification of the Töpfer algorithm for the minimization of a convex functional on a linear manifold[3]. Their algorithm is no more recursive and reaches a prescribed precision in a finite number of steps.

All the above mentioned methods consider only approximation problems in real normed linear spaces. We propose a generalization of the Remes algorithm for the linear approximation problem in real or complex normed linear spaces: We construct a sequence of discrete Tchebycheff problems, the strict approximations of which have an accumulation point which is a best approximation of our original problem.

2. THE APPROXIMATION PROBLEM

Let E be a normed linear space over the field $K = \mathbb{R}$ or $K = \mathbb{C}$ with norm $\|\cdot\|$, $G = [g_1,\dots,g_n]$ a n-dimensional subspace of E spanned by $g_1,\dots,g_n$. We want to find for a given element $f \in E$ an element $\tilde{g} \in G$, such that

$$d_G(f) := \min_{g \in G} \|f-g\| = \|f-\tilde{g}\| \quad .$$

If S_{E^*} denotes the unit cell of the dual space E^*, it is known that

$$\|h\| = \max_{L \in S_{E^*}} |L(h)| \quad \text{for any} \quad h \in E \quad .$$

We denote by $\text{ep}\, S_{E^*}$ the extremal points of S_{E^*}, fix m elements $L_1,\dots,L_m$ in $\text{ep}\, S_{E^*}$ and define for $j = 1,2,\dots,m$:

$$a_{j,k} := \text{Re}\, L_j(g_k) \quad \text{for} \quad k = 1,\dots,n \quad ,$$

$$a_{j,k+n} := -\,\text{Im}\, L_j(g_k) \quad \text{for} \quad k = 1,\dots,n \text{ and } K = \mathbb{C} \quad ,$$

$$f_j := \text{Re}\, L_j(f) \quad .$$

We consider for $g \in G$ the representation

$$g = \sum_{k=1}^{n} (x_k + i.x_{k+n})g_k$$

with $x_k \in \mathbb{R}$ for $1 \le k \le 2n$ and $x_{k+n} = 0$ for $K = \mathbb{R}$. With $r = n$ for $K = \mathbb{R}$ (resp. $r = 2n$ for $K = \mathbb{C}$) we get

$$\operatorname{Re} J_j(g-f) = \sum_{k=1}^{r} a_{j,k}x_k - f_j \quad .$$

Now we consider the discrete Tchebycheff problem in $\mathbb{R}^r$: Minimize

$$\max_{1\le j\le m} \sum_{k=1}^{r} |a_{j,k}x_k - f_j| = \max_{1\le j\le m} |\operatorname{Re} L_j(f-g)| \tag{2.1}$$

with respect to $x = (x_1,\ldots,x_r) \in \mathbb{R}^r$. For the minimal distance ρ of this problem we get naturally $\rho \le d_G(f)$. We will now define an algorithm consisting of such discrete Tchebycheff problems, the minimal distances of which will approximate the minimal distance of our original problem.

3. DISCRETE TCHEBYCHEFF APPROXIMATION

For abbreviation let

$$a_j := (a_{j,1}, a_{j,2}, \ldots, a_{j,r}) \quad ,$$
$$A := \{(a_j;f_j) \mid j = 1,2,\ldots,m\} \subset \mathbb{R}^r \times \mathbb{R} \quad .$$

Moreover let V be a subspace of $\mathbb{R}^r$ and W a linear manifold in $\mathbb{R}^r$ with difference space V. Let us denote by $\langle,\rangle$ the scalar product in $\mathbb{R}^r$.

Generalizing the concept of reference (Stiefel[7], Carasso and Laurent[3]), we consider

Definition 1 A subset $R = \{(\tilde{a}_i;\tilde{f}_i) \mid 1\le i\le k+1\} \subset A$ is called a *V-reference* if

(1) there exist $\lambda_i \in \mathbb{R}$ $(1\le i\le k+1)$ such that $\sum_{i=1}^{k+1} \lambda_i \tilde{a}_i \in V^{\perp}$,

(2) R is minimal, i.e. there exists no proper subset of R such that (1) holds.

Definition 2 $y \in W$ is called a *best approximation of the V-reference* R *in* W if

$$\inf_{x \in W} \max_{1 \le i \le k+1} |\langle \tilde{a}_i, x\rangle - \tilde{f}_i| = \max_{1 \le i \le k+1} |\langle \tilde{a}_i, y\rangle - \tilde{f}_i| \quad .$$

For constructing such a best approximation we define

$$h_R := \sum_{i=1}^{k+1} \lambda_i (\tilde{f}_i - \langle a_i, z\rangle) \Big/ \sum_{i=1}^{k+1} |\lambda_i|$$

for a fixed $z \in W$ and solve the linear system

$$\langle \tilde{a}_i, y\rangle = \tilde{f}_i - \langle \tilde{a}_i, z\rangle - h_R \cdot \operatorname{sgn} \lambda_i \qquad (i = 1,2,\dots,k+1)$$

with $y \in V$. Then $z+y$ is a best approximation of the V-reference R in W. The number $|h_R|$ is called the deviation of the V-reference R in W.

Now we consider the following construction: Let $R_1 \subset A$ be a $\mathbb{R}^r$-reference,

W_1 := {best approximations of the $\mathbb{R}^r$-reference R_1 in $\mathbb{R}^r$},

h_1 := deviation of R_1 in $\mathbb{R}^r$,

$V_1 := [pR_1]^{\perp} \cap V_o$.

(p is the projection of $\mathbb{R}^r \times \mathbb{R}$ into $\mathbb{R}^r$, $[M]$ the subspace generated by M in $\mathbb{R}^r$, $V_o := \mathbb{R}^r$.) Then V_1 is the difference space of the linear manifold W_1 and $\dim V_1 = r-k_1$, if R_1 has k_1+1 elements.

Now let $R_2 \subset A_1 := A-R_1$ be a V_1-reference in W_1 ,

W_2 := {best approximations of the V_1-reference R_2 in W_1 },

h_2 := deviation of R_2 in W_1 , $V_2 := [pR_2]^{\perp} \cap V_1$.

Then V_2 is the difference space of the linear manifold W_2 and $\dim V_2 := r-k_1-k_2$, if R_2 has k_2+1 elements. In this way we get a finite number of references $R_1, R_2, \ldots, R_s$ such that $\mathcal{R} = (R_1, R_2, \ldots, R_s)$ is a chain of references by

Definition 3 $\mathcal{R} = (R_1, R_2, \ldots, R_s)$ is called a *chain of references in* A if one of the following conditions holds:

(a) $V_s = \{0\}$,

(b) $A_s := A - \bigcup_{j=1}^{s} R_j = \phi$,

(c) pA_s is linear independent modulo $V_s^{\perp}$.

Therefore to each chain of references $\mathcal{R} = (R_1, R_2, \ldots, R_s)$ there are linear manifolds $W_1 \supset W_2 \supset \ldots \supset W_s$ with associated difference spaces $V_1 \supset V_2 \supset \ldots \supset V_s$ and a *deviation vector* $(h_1, h_2, \ldots, h_s)$. Each h_j is the deviation of the V_{j-1}-reference R_j in W_{j-1} .

Definition 4 $y \in W_s$ is called a best approximation of the chain $\mathcal{R} = (R_1, R_2, \ldots, R_s)$ if

$$\min_{x \in W_s} \max_{(a;f) \in A_s} |\langle a,x \rangle - f| = \max_{(a;f) \in A_s} |\langle a,y \rangle - f| .$$

If the chain $\mathcal{R}$ is defined by condition (a), then W_s has only one point, the best approximation of $\mathcal{R}$. In the case (b) all points of W_s are best approximations of $\mathcal{R}$, in the case (c) the linear system

$$\langle a,y \rangle = f \quad ((a;f) \in A_s)$$

is solvable with $y \in W_s$ and any solution is a best

approximation of the chain $\mathcal{R}$.

Definition 5 A chain of references $\mathcal{R} = (R_1, R_2, \ldots, R_s)$ is *regular* if each R_j with $j \geq 1$ has at least two elements.

We can get a regular chain by cancelling each R_j having only one element. Thereby the set of best approximations of the chain is not changed.

Definition 6 A chain $\mathcal{R} = (R_1, R_2, \ldots, R_s)$ is *strict* if the deviation vector $(h_1, h_2, \ldots, h_s)$ satisfies

$$h_j \geq h_{j+1} \quad \text{for} \quad j = 1,2,\ldots,s-1 \quad .$$

Definition 7 A point $z \in \mathbb{R}^r$ is called a *strict approximation* of the problem (2.1) if z is a best approximation of a strict chain of references $\mathcal{R} = (R_1, R_2, \ldots, R_s)$ with deviation vector $(h_1, h_2, \ldots, h_s)$ and

$$\max_{(a;f) \in A_s} |\langle a, z\rangle - f| \leq h_s \quad .$$

We can find such a strict approximation in a finite number of steps using the following:

Exhange Theorem (Carasso, Laurent[3]) Let V be a subspace of $\mathbb{R}^r$, W a linear manifold with difference space V, R_o a V-reference with deviation h_o in W. Moreover let R_1 be a V_o-reference for $V_o = [pR_o]^{\perp} \cap V$ with deviation h_1 in W_1 (W_1 denotes the set of best approximations of the V-reference R_o in W). Then there exists a subset $C_o \neq \phi$ of R_o such that $\tilde{R}_o = R_1 \cap (R_o - C_o)$ is a V-reference and C_o is a $\tilde{V}_o$-reference for $\tilde{V}_o = [p\tilde{R}_o]^{\perp} \cap V$. The deviation $\tilde{h}_o$ of $\tilde{R}_o$ in W satisfies

$$\tilde{h}_o = \alpha \cdot h_1 + (1-\alpha) \cdot h_o \quad \text{with} \quad 0 < \alpha \le 1 \quad .$$

4. THE ALGORITHM

We start our algorithm with a set $A^1 = A$ as in the foregoing sections. Without loss of generality we may assume that there exists a chain of references in A^1 .

In the 1-th step of the algorithm we have a discrete Tchebycheff problem (2.1) with $A = A^1$, a strict approximation

$$x^1 = (x_1^1, x_2^1, \ldots, x_r^1) \in \mathbb{R}^r$$

to this problem with an associated strict chain of references $\mathcal{R}^1 = (R_1^1, R_2^1, \ldots, R_{s_1}^1)$, difference spaces $V_1^1 \supset V_2^1 \supset \ldots \supset V_{s_1}^1$ and the deviation vector $(h_1^1, h_2^1, \ldots, h_{s_1}^1)$. Defining

$$g^1 := \sum_{k=1}^{n} (x_k^1 + i \cdot x_{k+n}^1) g_k$$

we determine $L \in \text{ep } S_{E^*}$ such that $L(f-g^1) = \|f-g^1\|$.
Now let us abbreviate

$$b_k^1 \quad := \quad \text{Re } L(g_k) \qquad \text{for} \quad k = 1,\ldots,n \quad ,$$

$$b_{k+n}^1 := - \text{Im } L(g_k) \qquad \text{for} \quad k = 1,\ldots,n \ \text{and} \ K = \mathbb{C} \quad ,$$

$$f^1 \quad := \quad \text{Re } L(f) \quad ,$$

$$b^1 \quad := \quad (b_1^1, b_2^1, \ldots, b_r^1) \quad .$$

If $b^1 = 0$ or $L(f-g_1) = h_1^1$, then g^1 is a best approximation to our problem and the algorithm stops. Otherwise we consider the conditions (a)-(c) of the definition 3 of $\mathcal{R}^1$. In the case (a) the set $\{(b^1;f^1)\}$ is a $V_{s_1}^1$-reference. Using the exchange theorem for $R_o := R_{s_1}^1$ and $R_1 := \{(b^1;f^1)\}$, we get a new $V^l_{s_l-1}$-reference $\tilde{R}_{s_l}$ such that

(1) $\{(b^l;f^l)\} \subset \tilde{R}_{s_l}$,

(2) $\tilde{R}_{s_l+1} := R^l_{s_l} - (R^l_{s_l} \cap \tilde{R}_{s_l}) = \phi$.

Finally we define

$$A^{l+1} := \begin{cases} \bigcup_{j=1}^{s_l-1} R^l_j \cup \tilde{R}_{s_l} , & \text{if } \tilde{R}_{s_l+1} \text{ has only one element} \\ A^l \quad \{(b^l;f^l)\} & \text{otherwise} \end{cases} .$$

We remark that the deviation vector is monotonically increasing in the lexicographic sense from step to step. For the convergence of the algorithm the following theorem holds.

THEOREM If the algorithm does not stop, we have:

(1) $\lim_{l\to\infty} h^l_l = d_G(f)$,

(2) there exists a subsequence of $\{g^l\}$, which converges to a best approximation of f with respect to G .

For the case $K = \mathbb{R}$, Carasso and Laurent[3] have given a similar method last year. However they use only a chain of references, not strict approximations. Therefore the description of their algorithm is a little more complicated and they had to decide a priori up to which precision the calculation of a best approximation is needed.

If $K = \mathbb{R}$ and G fulfils the Haar condition, then the algorithm is just the Remes algorithm[4]. For the uniform simultaneous approximation of functions by elements of Haar subspaces the algorithm is reduced to the method given in [1].

REFERENCES

1. Blatt, H.-P. (1975). Zur Konstruktion einer Minimallösung bei linearer Simultanapproximation, ISBN 26, Birkäuser Verlag, Basel and Stuttgart, 9-27.

2. Carasso, C. and Laurent, P.J. (1973). "Un algorithme pour la minimisation d'une fonctionnelle convexe sur une variété affine", Séminaire d'analyse numérique, Grenoble.

3. Carasso, C. and Laurent, P.J. (1976). "Un algorithme de minimisation en chaine en optimisation convexe", Séminaire d'analyse numérique, 242, Grenoble.

4. Laurent, P.J. (1967). Théorèmes de caractérisation d'une meilleure approximation dans un espace normé et généralisation de l'algorithme de Rémès, *Num. Math.* 10, 190-208.

5. Rémès, E. (1934). Sur le calcul effectif des polynomes d'approximation de Tchebycheff, *C.R. Acad. Sci. Paris* 199, 337-340.

6. Rice, J.R. (1963). Tchebycheff Approximation in Several Variables, *Trans. Amer. Math. Soc.* 109, 444-466.

7. Stiefel, E.L. (1959). Über diskrete und lineare Tsceby-scheff-Approximationen, *Num. Math.* 1, 1-28.

8. Töpfer, H.J. (1967). Tschebyscheff Approximation und Austauschverfahren bei nicht erfüllter Haarscher Bedingung, ISM 7, Birkhäuser Verlag, Basel and Stuttgart, 71-89.

POLYNOMIAL APPROXIMATION OF SHAPES BASED ON CONDENSED REPRESENTATIONS OF BERNSTEIN POLYNOMIALS IN ONE AND SEVERAL VARIABLES

E.L. Ortiz and M.R.J. da Silva

University of London
Imperial College, Mathematics Department
London, England

1. INTRODUCTION

Bernstein polynomials enjoy the remarkable property of imitating quite closely the global behaviour of the continuous functions with which they are associated. However, their slow speed of convergence puts a serious limitation to their applicability in numerical approximation problems.

P.J. Davis[1], and since then many authors, has pointed out the possibility of using Bernstein approximants when shape is more important than a close point-to-point approximation. Even so, fairly high degree approximants seem to be required in most situations.

Let us assume that, in the process of approximating to the *shape* of a given curve, there is given a tolerance parameter ε related to the accuracy to which variations in shape cease to be detectable or relevant for the problem in hand. In this paper we show that under fairly weak smoothness conditions on the original function f, it is possible to apply the process of *numerical condensation* to an initial Bernstein approximant of high degree and reduce it quite considerably, keeping its shape approximating properties only slightly changed. A ripple will appear on the original approximating curve, the amplitude

2ε of which is known in advance.

On the lines of Lanczos[3], Ortiz[6] and Helman-Ortiz[2], we derive sufficient conditions for the possibility of such numerical condensation process, in terms of the givven error bound ε, for Bernstein approximants in one and two dimensions. Finally, an example is given on condensed shape approximation of a polygonal function by means of a single polynomial of fairly low degree.

2. BERNSTEIN POLYNOMIALS AND NUMERICAL CONDENSATION

Let

$$p(x) = b_n + b_{n-1}x + \ldots + b_o x^n \in \mathcal{P}_n$$

be a polynomial of degree n in x, let us assume that we are interested in the values of p for x in a compact interval $J = [a,b]$, that $\varepsilon > 0$ is a given error bound for the values of $p(x)$ in J, and that those values are actually computed by means of a backwards recurrence multiplication scheme of the form

$$p(x) = p_o \begin{cases} p_{n+1} = 0 \\ p_k = xp_{k+1} + b_k \quad , \quad k = 0(1)n \end{cases} \quad . \tag{2.1}$$

(see Luke[5]). Clearly, other multiplication schemes could be introduced here.

We say that a polynomial p_ε is a condensed representation of p in J if

$$\|p-p_\varepsilon\| < \varepsilon$$

(where $\|\cdot\|$ stands for the usual norm of $C[J]$), and the computational effort required to compute values of p_ε is, in some sense (see Ortiz[6]) smaller than that required to compute values of p. In the present case we shall measure the computational effort by the number of multiplications required to compute values of p (or p_ε) by means of scheme (2.1) or its multidimensional extensions (see [6]).

In the case of polynomials of one independent variable it is easy to see that condensation is possible if and only if

$$|b_o| < 2^{2n-1}\varepsilon \quad .$$

A sufficient condition for the existence of a polynomial $p_\varepsilon \in P_{n-r-1}$, $r \geq 0$, is

$$\sum_{i=0}^{r} 4^i \, |a_{n-i}| \sum_{j=0}^{r-i} \binom{2(n-i)}{j} < 2^{2n-1}\varepsilon \quad . \tag{2.2}$$

Let $P(x,y) \in P_{n+m} \equiv \text{span}\{x^i y^j\}_{i=0,j=0}^{n\ ,m}$, then we write

$$P(x,y) = \sum_{i=0}^{n} \sum_{j=0}^{m} a_{ij} x^i y^j - V_{xn}^T \cdot \underline{A} \cdot V_{ym} \quad ,$$

where $V_{tr} = (t^0, t^1, \ldots, t^r)^T$ and $\underline{A} = ((a_{ij}))$, $i = 0(1)n$, $j = 0(1)m$. Let us assume further that we are interested in the values of P over the square domain $Q = \{(x,y) \in \mathbb{R}^2 : -1 \leq x,y \leq 1\}$.

The notion of numerical condensation in several variables is discussed in Ortiz[6] where, in particular, the following sufficient conditions for that process to be possible in square domains are given:

i) $$\sum_{s=0}^{n} \sum_{k=0}^{n-s} |a_{n-s\ m}| < 2^{2n-1}\varepsilon \quad ,$$

if all the coefficients of $\underline{A}$ are relevant, or

ii) $\sum_{i=0}^{n} |a_{n-i\ i}| < 4^{n-1}\varepsilon$,

if the entries of $\underline{A}$ are equal to zero for $i+j > n$, which is the case of polynomials P obtained by truncation to degree n in x and y of a convergent power series expansion in two variables.

These conditions are easily extended, by recursion, to the case of several independent variables.

3. BERNSTEIN POLYNOMIALS: SUFFICIENT CONDITIONS FOR NUMERICAL CONDENSATION

The Bernstein operator $B_n: C[0,1] \to P_n$ is given by

$$B_n(x,f) = \sum_{k=0}^{n} f_k q_{nk}(x) \quad ,$$

where

$$f_k = f(k/n) \quad \text{and} \quad q_{nk}(x) = \binom{n}{k} x^k (1-x)^{n-k} \quad .$$

$B_n(x,f)$ can also be written in conventional polynomial form as

$$B_n(x,f) = \sum_{k=0}^{n} \binom{n}{k} \Delta^k f_o x^k \quad . \tag{3.1}$$

Remark 1.3 We notice that B_n maps the *class* of functions taking on the values f_k at the mesh points $\{k/n\}$, $k = 0(1)n$, into one polynomial. In particular, if $\hat{p}(x) \in P_n$ is the polynomial which interpolates the table $\{k/n, f_k\}_{k=0}^{n}$, we have

$$B_n(x,f) \equiv B_n(x,\hat{p}) \quad .$$

As a result, conditions under which $B_n(x,f)$ may be condensed can always be stated in terms of $\hat{p}$.

Let us discuss now sufficient conditions for the feasibility of the condensation process:

i) In our case $b_{n-i} = \binom{n}{i}\Delta^{n-i}f_o$, $i = 0(1)n$, then condition (2.2) becomes

$$\sum_{i=0}^{r} 4^i \binom{n}{i} |\Delta^{n-i} f_o| \sum_{j=0}^{r-i} \binom{2(n-i)}{j} < 2^{2n-1}\varepsilon \quad .$$

The condensation process from P_n to P_{n-1} requires that the following smoothness condition, in terms of the differences of $f(x)$, be satisfied:

$$|\Delta^n f_o| < 2^{2n-1}\varepsilon \quad .$$

ii) Making use of Lagrange's representation of $\hat{p}(x) = \sum_{j=0}^{n} \hat{p}_j x^j$, we find that

$$B_n(x,f) = \sum_{j=0}^{n} \hat{p}_j B_n(x,x^j) = \sum_{i=0}^{n} \left(\sum_{j=i}^{n} a_{ij}\hat{p}_j\right)x^i \quad , \tag{3.2}$$

with

$$a_{ij} = \binom{n}{i} \Delta^i x^j|_{x=0} = \lambda_i n^{i-j}\sigma_{ij} \quad ,$$

where $\lambda_i = i!\binom{n}{i}/n^i$, $i = 0(1)n$, and σ_{ij} , $i,j = 0(1)n$, are the Stirling numbers of the second kind.

The matrix $\underline{A}$ is upper triangular with non-negative elements. Furthermore, it is column stochastic, since $\sum_{i=0}^{n} a_{ij} = 1$ for any $j = 1(1)n$, as it follows from the fact that $B_n(1,x^j) = 1$.

Equating coefficients of like powers of x in (3.1) and (3.2) leads to

$$\binom{n}{i}|\Delta^i f_o| = \sum_{j=i}^{n} a_{ij}\hat{p}_j \quad , \quad i - 0(1)n \quad .$$

Thus, for $i = n$: $\Delta^n f_o = \lambda_n \hat{p}_n = n!\ p_n/n^n$, which gives a condition on the smoothness of f for the possibility to condense $B_n(x,f)$ from P_n to P_{n-1} .

In terms of the leading coefficient of $B_n(x,f)$ we get the condition

$$|\hat{p}_n| < \frac{n^n 2^{2n-1}}{n!} \varepsilon \quad .$$

Besides, since

$$\binom{n}{i}|\Delta^i f_o| \sum_{j=i}^{n} a_{ij}|\hat{p}_j| = m||\underline{A}||_\infty$$

(where $||\cdot||_\infty$ is the infinity norm and $m = \max_{0\le j\le n}|\hat{p}_j|$), an r-step condensation process, i.e. from P_n to P_{n-r} , $r \ge 1$, requires that

$$\sum_{i=0}^{r} 4^i \sum_{j=0}^{r-i} \binom{2(n-i)}{j} < \frac{2^{2n-1}}{m||A||_\infty} \varepsilon < \frac{2^{2(n-1)}}{m} \varepsilon$$

be satisfied.

iii) Let us define $f^T = (f_k)$, $Q^T = (q_{nk})$, $T^{*T} = (T_k^*(x))$, where $k = 0(1)n$. If $\underline{C} = ((c_{ij}))$ is the lower triangular matrix with elements $c_{ij} = (-1)^{i-j}\binom{n}{i}\binom{i}{j}$ for $n \ge i \ge j \ge 0$ and zero otherwise, and $\underline{D} = ((d_{ij}))$ is the upper triangular matrix with elements $2^{2j}d_{ij} = \binom{2j}{j}$, $2^{2j}d_{ij} = 2\binom{2j}{j-i}$ for $0 \le i \le j \le n$ and zero otherwise, we can write

$$Q^T = V_{xn}^T \cdot \underline{C} \quad \text{and} \quad V_{xn} = T^{*T} \cdot \underline{D} \quad .$$

It then follows that

$$B_n(x,f) = Q^T.f = V_{xn}^T.\underline{C}.f = T^{*T}.\underline{G}.f \quad ,$$

with $\underline{G} = \underline{D}.\underline{C}$.

If $\hat{p}^T = (\hat{p}_k)$, $k = 0(1)n$, then $f = \underline{U}.\hat{p}$, where $\underline{U} = ((u_{ik}))$, $i,k = 0(1)n$, $u_{oo} = 1$ and $u_{ik} = (i/n)^k$ otherwise. Since $\underline{C}.\underline{U} = \underline{A}$, we have, with $H = ((h_{ij})) = \underline{D}.\underline{A}$, $i,j = 0(1)n$:

$$B_n(x,f) = T^{*T}.\underline{G}.f = T^{*T}.\underline{H}.\hat{p} \quad .$$

A sufficient condition for the feasibility of r steps of the condensation process is, then

$$\sum_{i=0}^{r} \sum_{j=0}^{i} h_{n-i\ n-j} < \varepsilon/m \quad .$$

Remark 2.3 Let $f \in C[0,1]$ be a convex function in $[0,1]$. We say that a function $g \in C[0,1]$ is an ε-convex approximation to f provided that $||f-g|| < \varepsilon$.

It is well known that if $f \in C[0,1]$ is convex in $[0,1]$, so is $B_n(x,f)$. If $p_\varepsilon(x)$ is a condensed representation of $B_n(x,f)$, the convexity is no longer preserved but p_ε is ε-convex with respect to $B_n(x,f)$ and ε'-convex with respect to f , where $\varepsilon' = \varepsilon + ||f(x)-B_n(x,f)||$.

4. BIVARIATE BERNSTEIN POLYNOMIALS

Let $f = f(x,y)$ be a continuous function of two independent variables defined in $Q^* = \{(x,y) \in \mathbb{R}^2 : 0 \le x,y \le 1\}$. The Bernstein polynomial associated with f is defined as (see Lorentz[4])

$$B_{nm}(x,y,f) = \sum_{i=0}^{n} \sum_{j=0}^{m} f(\tfrac{i}{n},\tfrac{j}{m}) q_{ni}(x) q_{mj}(y) \in \mathcal{P}_{n+m} \quad ,$$

where the q's are defined as in section 3.

We have indicated that the conditions for condensation in one variable are easily expressible in terms of the smoothness of the function. We will now do the same for $f(x,y)$ (for details on the multi-dimensional extension of the condensation process see Ortiz[6]). As

$$f(x,y) \sim \sum_{i=0}^{n} \binom{n}{i} \Delta^{i}_{(x,h_1)} f(0,y) x^{i} \quad ,$$

with $h_1 = 1/n$, and

$$f(0,y) \sim \sum_{j=0}^{m} \binom{m}{j} \Delta^{j}_{(y,h_2)} f(0,0) y^{j} \quad ,$$

with $h_2 = 1/m$, it follows that

$$B_{nm}(x,y,f) = \sum_{i=0}^{n} \sum_{j=0}^{m} \binom{n}{i}\binom{m}{j} \Delta^{i}_{(x,h_1)} \Delta^{j}_{(y,h_2)} f(0,0) x^{i} y^{j} \quad .$$

Furthermore

$$\Delta^{j}_{(y,h_2)} \Delta^{i}_{(x,h_1)} f(0,0) = \Delta^{i}_{(x,h_1)} \Delta^{j}_{(y,h_2)} f(0,0)$$
$$= \sum_{r=0}^{i} \sum_{s=0}^{j} (-1)^{i+j-(r+s)} \binom{i}{r}\binom{j}{s} f(rh_1, sh_2) \quad .$$

Let $\underline{E} = ((e_{ij}))$, $i = 0(1)n$, $j = 0(1)m$, be the coefficient matrix of the polynomial $B_{nm}(x,y,f)$, then $e_{ij} = \binom{n}{i}\binom{m}{j} \Delta^{i}_{(x,h_1)} \Delta^{j}_{(y,h_2)} f(0,0)$. Since $V^{T}_{tn} = \underline{D}.T^{*T}_{tn}$, we have

$$B_{nm}(x,y,f) = V^{T}_{xn}.\underline{E}.V_{ym} = T^{*T}_{xn}.\underline{S}.T^{*T}_{ym} \quad , \tag{4.1}$$

with $\underline{S} = \underline{D}_1.\underline{E}.\underline{D}_2$ ($\underline{D}_1, \underline{D}_2$ are matrices of type $\underline{D}$ and dimension $(n+1)\times(n+1)$ and $(m+1)\times(m+1)$ respectively).

From (4.1)

$$\mathrm{coef}\{T^*_m(y)\} = \sum_{i=0}^{n} s_{im} T^*_i(x) \quad ,$$

besides

$$d_{im} = \sum_{r=0}^{n} \sum_{s=0}^{m} (\underline{D}_1)_{ir} e_{rs} (\underline{D}_2^T)_{sm} = 2^{1-2m} \sum_{r=i}^{n} (\underline{D}_1)_{ir} e_{rm}$$
$$= 2^{1-2m} \sum_{r=0}^{n-i} (\underline{D}_1)_{i\ i+r} e_{i+r\ m} = 2^{1-2m} \sum_{r=0}^{n-i} (\underline{D}_1)_{i\ n-r} e_{n-r\ m}$$
$$= 2^{2(1-(n+m))} \sum_{r=0}^{n-i} 2^{2r} \binom{2(n-r)}{n-r-i} e_{n-r\ m} \quad .$$

Thus,

$$\mathrm{coef}\{T^*_m(y)\} = 2^{2(1-(n+m))} \sum_{i=0}^{n} \sum_{r=0}^{n-i} 2^{2r} \binom{2(n-r)}{n-r-i} e_{n-r\ m} T^*_i(x)$$
$$= 2^{2(1-(n+m))} \sum_{r=0}^{n} 2^{2r} e_{n-r\ m} \sum_{i=0}^{n-r} \binom{2(n-r)}{n-r-i} T^*_i(x)$$
$$= 2^{2(1-(n+m))} \sum_{r=0}^{n} 2^{2r} e_{n-r\ m} \sum_{i=0}^{n-r} \binom{2(n-r)}{i} T^*_{n-r-i}(x) \ .$$

Therefore

$$\mathrm{coef}\{T^*_m(y)\} \le 2^{2(1-(n+m))} \sum_{r=0}^{n} 2^{2r} |e_{n-r\ m}| \left(\sum_{i=0}^{n-r}{}' \binom{2(n-r)}{i} \right)$$
$$= 2^{1-2m} \sum_{k=0}^{n} |e_{n-r\ m}| \quad .$$

Then, the condensation process for $B_{nm}(x,y,f)$ is possible if

$$\sum_{r=0}^{n} |e_{n-r\ m}| < 2^{2m-1}\varepsilon \quad .$$

A similar condensation condition holds true for the second index.

The sufficient smoothness conditions for the possibility of numerical condensation of $B_n(x,y,f)$ are then

$$\sum_{i=0}^{n} \binom{n}{i} \left|\Delta^{m}_{(y,h_2)} \Delta^{i}_{(x,h_1)} f(0,0)\right| < 2^{2m-1}\varepsilon \quad ,$$

and

$$\sum_{j=0}^{m} \binom{m}{j} \left|\Delta^{n}_{(x,h_1)} \Delta^{m}_{(y,h_2)} f(0,0)\right| < 2^{2n-1}\varepsilon \quad .$$

These conditions state that it is possible to project $B_{nm}(x,y,f)$ onto a proper subspace of P_{n+m} without introducing an error greater than ε in the numerical values taken by B_{nm} on Q^* .

For r steps of the condensation process we find the condition

$$\sum_{j=0}^{r} \left|\text{coef } T^*_{m-j}(y)\right| \leq \sum_{j=0}^{r} 2^{1-2(m-j)} \sum_{i=0}^{n} \left|e_{n-i\ m-j}\right| < \varepsilon \quad ;$$

thus,

$$\sum_{j=0}^{r} 4^{j} \sum_{i=0}^{m} \left|e_{n-i\ m-j}\right| < 2^{2m-1}\varepsilon \quad ,$$

and

$$\sum_{i=0}^{r} 4^{i} \sum_{j=0}^{m} \left|e_{n-1\ m-j}\right| < 2^{2n-1}\varepsilon \quad .$$

Remark 1.4 The smoothness of f along the x (or y) axis determines the possibility of numerical condensation in the variable y (or x) respectively.

Remark 2.4 Sufficient conditions for condensation in two variables can be stated as one-dimensional condensation conditions for the functions

$$\Delta^{m}_{(y,h_2)} f(0,0) \quad , \quad \Delta^{n}_{(x,h_1)} f(0,0) \quad .$$

5. NUMERICAL EXAMPLE

Bernstein polynomials are natural candidates for numerical condensation, as our sufficient conditions clearly show. For them "high frequency" components take on a very low amplitude value. The coefficient of $T^*_n(x)$ in $B_n(x,x^n)$ is equal to

$$\lambda_n/2^{2n-1} = \Big(\prod_{i=1}^{n-1} (1-i/n)\Big)/2^{2n-1} \quad ;$$

in particular, for $n = 5$ and 10 those coefficients are 7.5×10^{-5} and 3.5×10^{-10} respectively.

Let us consider the shape approximation problem of the polygonal function f - defined by the points $(0,0)$; $(.2,.6)$; $(.6,.8)$; $(.9,.7)$ and $(1,0)$ - by means of a *single polynomial* of a fairly low degree.

If we consider the approximant $B_{10}(x,f)$ and admissible error bounds $\varepsilon = 3.5\times10^{-3}$, $\varepsilon' = 7.0\times10^{3}$ and $\varepsilon'' = 2.5\times10^{-2}$, it is possible to condense B_{10} to polynomial representations of degree 6, 5 and 4 respectively without exceeding the given error bounds.

In each of Figures 1, 2 and 3 we show the graph of $f(x)$, $B_{10}(x,f)$, the condensed representation p_ε of $B_{10}(x,f)$ to degree r , and $B_r(x,f)$, for $r = 4$, 5 and 6 respectively. We can appreciate in Figure 2 that for $r = 5$ the adjustment between B_{10} and its condensed representation is fairly close.

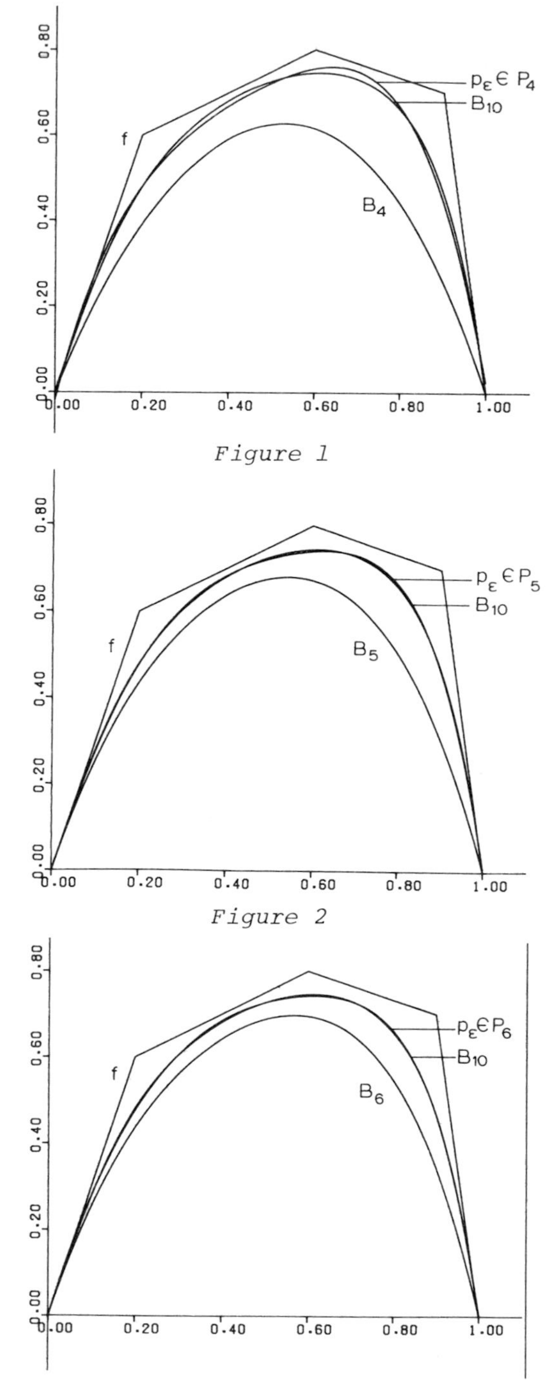

Figure 1

Figure 2

Figure 3

REFERENCES

1. Davis, P.J. (1963). "Interpolation and Approximation", Blaisdell Publishing Co., New York.

2. Helman, H. and Ortiz, E.L. The Method of Condensation (in preparation).

3. Lanczos, C. (1938). Trigonometric Interpolation of Empirical and Analytical Functions, *J. Math. Phys.* 17, 129-199.

4. Lorentz, G.G. (1953). "Bernstein Polynomials", University of Toronto Press, Toronto.

5. Luke, Y.L. (1969). "The Special Functions and Their Approximation", Vol.I, Academic Press, New York.

6. Ortiz, E.L. (1977). Polynomial Condensation in One and Several Variables, *in* "Topics in Numerical Analysis III" (Ed. J. Miller) Academic Press, New York, 73-106.

A PRACTICAL METHOD FOR ESTIMATING APPROXIMATION ERRORS IN SOBOLEV SPACES

Jean Meinguet

Institut de Mathématique pure et appliquée
Université de Louvain
Louvain-la-Neuve, Belgium

1. INTRODUCTION

This paper originates from some recent research work whose primary purpose was to devise a *method of practical value* for obtaining *realistic upper bounds of approximation errors* in a *wide variety of situations*. As shown in [4,5,6], such a method (see Section 4) can be elaborated from a *structural analysis of error coefficients*, which evolves quite naturally in the setting of operator theory in normed linear spaces while referring to such classical tools as the *Peano kernel theorem* and its generalization known as the *Bramble-Hilbert lemma*. As exemplified in [4,5,6] and also in [1,3] that method can yield at *reasonable cost* explicit upper bounds for certain *generic constants* which pervade the modern literature on error estimation (typically in connection with the rate of convergence of the finite element method, see e.g. [2], Theorems 2,4,5,6).

We will review and summarize here some of the significant results obtained so far in matter of *quantitative estimation* of interpolation and (more generally) approximation errors, the reader being naturally referred to the original papers[4,5,6] for a more complete theoretical or practical study of that extensive topic. More precisely, the underlying abstract analysis outlined in Section 2 is developed more fully, with explicit motivations, in [6] (specially in Sections 2, 3, 4). As for the two wide classes of (multivariate) applications to be considered in the first instance, namely *pointwise or uniform approximation in* $C^m(\bar{\Omega})$ and *mean-square approximation in* $H^m(\Omega)$, only the latter will be studied here (following closely Section 4 of [5]) in view of its novelty and its less elementary character (the former class is considered not only in [5] but also in [4,6]). Relevant *key estimates in Sobolev spaces* (see Section 3) are obtained by making use of a certain extension of the Taylor formula (with integral expression of the remainder), which proves both natural and convenient; in spite of apparent similarities, this far-reaching representation formula in $H^m(\Omega)$ essentially differs from the one exploited in [1,3] and, anyhow, it does not suffer at all from the intrinsic limitation emphasized in [1] (see Remark 2-3, p.16).

2. STRUCTURAL ASPECTS OF ERROR COEFFICIENTS

As explained in [4,5,6], we are basically concerned with *inequalities* of the form:

$$\|Rf\|_Z \leq c\|Uf\|_Y \quad , \quad \forall f \in X \quad , \tag{2.1}$$

where X, Y and Z are given linear spaces (over $\mathbb{R}$ or over $\mathbb{C}$), $\|\cdot\|_Y$ and $\|\cdot\|_Z$ denote norms given respectively on Y and Z, $U: X \to Y$ is a given *linear surjection* (i.e., $Y := U(X)$) and $R: X \to Z$ is a given *linear* mapping verifying the assumption

(H1) R is continuous with respect to U .

Condition (H1) is natural: it is indeed equivalent to the

requirement that (2.1) hold with c denoting some *finite* positive constant, so as to yield some genuine appraisal of Rf (regarded as *unknown*) in terms of Uf (supposed to be *known*). A problem of *practical* interest, specially perhaps in numerical analysis and in approximation theory, is to find *realistic upper bounds* c for the *theoretical error coefficient*

$$c_o := \sup_{\substack{f \in X \\ Uf \neq 0}} \|Rf\|_Z / \|Uf\|_Y \equiv \min\{c \in \mathbb{R} \ \text{verifying (2.1)}\} \quad . \tag{2.2}$$

As a matter of fact, it is quite exceptional that the value of c_o itself can be determined accurately at reasonable cost (in mean-square approximation problems, for example, it can be shown that c_o^{-2} is the lowest point of the spectrum of some operator); needless to say, this does not mean at all that the structure of such error coefficients is not worth studying carefully.

An interesting expression for c_o readily follows from the inclusion relation

$$K \equiv \text{kernel } U \subset \text{kernel } R \tag{2.3}$$

between linear subspaces of X, which is implied by (2.1) under (H1). According to a classical *factorization theorem* (borrowed essentially from elementary set theory), (2.3) is equivalent to

$$R = QU \quad , \tag{2.4}$$

where $Q: Y \to Z$ denotes some uniquely defined linear mapping; moreover, Q is surjective (i.e., onto) iff R is surjective, injective (i.e., one-to-one) iff $K \equiv \text{kernel } R$. In actual fact,

Q has the explicit expression

$$Q = RV \quad , \tag{2.5a}$$

where $V: Y \to X$ denotes an arbitrary *right inverse* of U, i.e., a (possibly nonlinear) mapping such that

$$UV = 1_Y \quad , \tag{2.6}$$

where 1_Y denotes the identify mapping of Y (the existence of such V's is known to be equivalent to the surjectivity of U). Hence it follows that c_o, originally defined by (2.2), is explicitly given by

$$c_o = \|Q\| \equiv \|RV\| \quad , \quad \forall V: Y \to X \text{ such that } UV = 1_Y \, . \tag{2.7}$$

For various reasons, both theoretical and practical, it proves useful to introduce certain additional spaces and mappings, as suggested by the so-called *canonical decomposition of mappings*. According to that corollary of the factorization theorem recalled above, every mapping T of an arbitrary set E into an arbitrary set F can be represented (in mnemonic notations) by the following diagram:

$$E \xrightarrow{T_{cs}} E/\mathcal{T} \xrightarrow{T_b} T(E) \xrightarrow{T_{ci}} F \quad ; \tag{2.8}$$

here $\mathcal{T}$ denotes the equivalence relation in E that is associated with T (i.e., the relation "$e_1 \in E$, $e_2 \in E$, $Te_1 = Te_2$"), T_{cs} is the *canonical surjection* of E onto the quotient set $E/\mathcal{T}$, T_b is the *bijection associated with* T, T_{ci} is the *canonical injection* of $T(E)$ into F. We are thus led eventually to the following *commutative diagram*:

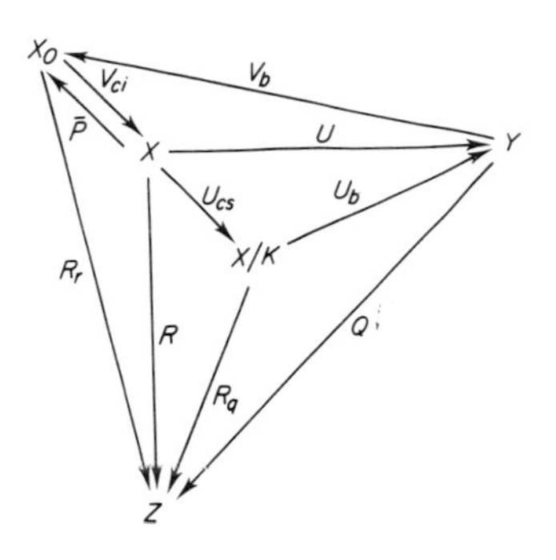

Figure 1

where the subset of X defined by

$$X_o := V(Y) \equiv VU(X) \tag{2.9}$$

is a linear space iff $V: Y \to X$ is linear, X/K is nothing else than the linear space $X/\mathcal{U}$, R_r is the *restriction* of R to X_o, so that

$$Q = R_r V_b \quad \text{with} \quad R_r = RV_{ci} \quad , \tag{2.5b}$$

R_q is the *quotient mapping* that results from R (always supposed to be compatible with the equivalence relation $\mathcal{U}$ in X) by factorization through U_{cs}, so that

$$Q = R_q U_b^{-1} \quad \text{with} \quad R_q = RVU_b \quad . \tag{2.5c}$$

As for the mapping $\bar{P}$ (and its 'complement' P) defined by

$$\bar{P} \equiv 1_X - P := VU \quad \text{with} \quad UV = 1_Y \quad , \tag{2.10}$$

they are both *idempotent* under the only condition that V be *homogeneous*, in which case they are homogeneous too while the range of P is the whole of K. Whenever V is a *linear* right inverse of U, which most frequently occurs in practice, $\bar{P}$ (resp. P) is a *linear projector* of X *onto* X_o (resp. K), so that

$$X = K \oplus X_o \quad ; \tag{2.11}$$

this *direct sum decomposition* of X is equivalent to the following *representation formula* in X :

$$f = Pf + VUf \ , \qquad \forall f \in X \ , \tag{2.12}$$

where Pf is to be regarded as an *approximant* of f and VUf as the associated *remainder*.

Except when Y and Z are inner product spaces, where direct variational techniques may prove preferable, it seems that the actual determination of c_o can only be based on some explicit representation of $Q: Y \to Z$. Such a preliminary result can be obtained by resorting to suitable *Peano kernel theorems*, whose abstract formulation reduces indeed to the trivial identity

$$Rf \equiv RVUf \ , \qquad \forall f \in X \quad \text{with} \quad UV = 1_Y \ . \tag{2.13}$$

Even though *accessible standard forms* for Rf can be found (in terms of Uf), there remains the problem of *calculating* $\|Q\|$; this often proves surprisingly difficult, in view of the complexity of the possible 'resonances' between *functions* Uf and *kernels* RV.

As a general rule, *the data* X, Y, U *may be regarded as essentially fixed whereas* R *is essentially variable*: typically, U is a differential operator over some classical functional space (such as $C^m(\bar{\Omega})$ or $H^m(\Omega)$) whereas R is the *error* associated with some specific approximation process. In view of this fundamental distinction between the data, it is by far more justified to concentrate, for *standard* choices of fixed data, on a realistic minimization of the positive constant d (considered as *independent* of R) in

$$\|Rf\|_Z \le \|R\| d \|Uf\|_Y \quad , \qquad \forall f \in X \quad , \tag{2.14}$$

than to attempt to cope with the *specific* difficulties inherent in the precise evaluation of $\|R\|$ in (2.14) or of $c_o \equiv \|RV\|$ in (2.1). For these reasons, we deem it justified to qualify as *practical* the estimates of the form (2.14) and the associated *error coefficient*

$$d_o := \inf\{d \in \mathbb{R} \text{ verifying (2.14)}, \ \forall R \in L(X;Z)\} \quad ; \tag{2.15}$$

of course, for this to make sense at all, the assumption:

(H2) X can be normed so that R: X → Z is continuous and there exists a bounded right inverse V of U: X → Y ,

must be added to (H1). As easily seen, d_o can be regarded dually, either as the solution of the *maximum-minimum* problem

$$d_o = \|U_b^{-1}\| = \sup_f \inf_{k \in K} \|f-k\|_X = \sup_f \inf_V \|VUf\|_X \quad , \tag{2.16a}$$

or as the solution of the *minimum-maximum* problem

$$d_o = \inf_V \|V\| = \inf_V \sup_f \|VUf\|_X \quad , \tag{2.16b}$$

f ranging in X over the semisphere $\|Uf\|_Y = 1$ and V ranging over the set of *homogeneous*, bounded right inverses of U . As a matter of fact, (2.14) with $d := \|V\|$ can be regarded as following from either of the two *optimal inequalities*

$$\|Rf\|_Z \leq \|R\| \inf_{k\in K} \|f-k\|_X \equiv \|R\| \; \|f\|_{X/K}, \quad \forall f \in X \quad , \tag{2.17a}$$

$$\|Rf\|_Z \leq \|RV\| \; \|Uf\|_Y \quad , \qquad \forall f \in X \quad , \tag{2.17b}$$

which, as a rule, cannot be compared on a common basis.

It must be emphasized that various *conditional equivalences between norms* have to play a prominent role in practical estimation problems, namely those resulting from the sharp appraisals (see the commutative diagram above):

- *if* U *is continuous, then*

$$\|Uf\|_Y \leq \|U\| \; \|f\|_{X/K} \quad , \tag{2.18a}$$

$$\|Uf\|_Y \leq \|U\| \; \|VUf\|_X \quad ; \tag{2.19a}$$

- *if* V *is any (homogeneous) bounded right inverse of* U *, then*

$$\|f\|_{X/K} \leq \|U_b^{-1}\| \; \|Uf\|_Y \leq \|V\| \; \|Uf\|_Y \quad , \tag{2.18b}$$

$$\|VUf\|_X \leq \|V\| \; \|Uf\|_Y \quad ; \tag{2.19b}$$

- *if the (homogeneous) composite* VU *is bounded, then*

$$\|VUf\|_X \leq \|VU\| \; \|f\|_{X/K} \quad , \tag{2.20a}$$

while, at any rate,

$$\|f\|_{X/K} \leq \|VUf\|_X \quad . \tag{2.20b}$$

These inequalities can be regarded as abstract generalizations of the classical *Bramble-Hilbert lemma* (which essentially asserts the equivalence over $H^m(\Omega)/\mathcal{P}_{m-1}$ of the quotient norm and of the Sobolev seminorm of order m), U and U_b^{-1} being then both continuous. It should also be noticed that, in view of (2.10), searching for a right inverse of minimal norm V_o of U amounts to searching for a *best-approximation projector* P_o of X onto K ; such optimal mappings always exist whenever K is finite-dimensional but may be nonlinear (though homogeneous, as exemplified by the well-known Chebyshev operator).

3. KEY ESTIMATES IN MULTIVARIATE MEAN-SQUARE APPROXIMATION

From the foregoing structural analysis, it is fairly easy to elaborate a general method for finding realistic upper bounds of theoretical (resp. practical) error coefficients c_o (resp. d_o) ; described and illustrated briefly in Section 4, that method essentially makes use of so-called *key estimates*. Roughly speaking, most of the relevant quantitative estimation problems are typically concerned either with *pointwise approximation* or with *mean-square approximation*. Of these two wide classes of significant applications, only the latter will be considered here, following closely [5] (for the former class, see e.g. [6], Section 5).

With n any positive integer, let Ω be a (non-empty) *bounded open convex* subset of R^n, of (Euclidean) diameter h and of Lebesgue measure S ; as a matter of fact, these mild restrictions concerning Ω could be easily weakened by substituting measures of suitable compact support for the Lebesgue measure throughout the following. In matter of mean-square approximation, it is both natural and convenient to take as *fixed standard data* X and U, respectively:

- the *Sobolev space* $H^m(\Omega)$ (with m integer ≥ 1) of (classes of) numerical functions $f \in L_2(\Omega)$ for which all partial derivatives $\partial^\alpha f$ (in the distributional sense) with $|\alpha| \leq m$ also belong to $L_2(\Omega)$ (we use here the classical multi-index notations); equipped with the family of seminorms

$$|f|_j = \Big(\sum_{i_1,\ldots,i_j=1}^{n} \int_\Omega \Big| \frac{\partial^j f(x)}{\partial x_{i_1} \ldots \partial x_{i_j}} \Big|^2 dx \Big)^{1/2} , \qquad 0 \leq j \leq m , \tag{3.1a}$$

equivalent to the norm

$$\|f\|_m = \Big(\sum_{j=0}^{m} |f|_j^2 \Big)^{1/2} , \tag{3.1b}$$

$H^m(\Omega)$ is a Hilbert space; it can be defined equivalently as the completion with respect to the norm (3.1b) of the linear space $C^m(\bar{\Omega})$ of the numerical functions which are *uniformly continuous* in Ω together with all their partial derivatives of order $\leq m$.

- the m-th *total derivative* operator D^m, the resulting partial derivatives of order m being all to be interpreted in the distributional sense; since Ω is connected, the kernel K of U in X is known to be simply the linear space P_{m-1} of polynomials over R^n of total degree $\leq m-1$; as for the space $Y := D^m(H^m(\Omega))$, equipped with the natural norm

$$||D^m f||_Y = |f|_m \tag{3.2}$$

defined by (3.1a), it can be identified by the (non-isometric!) isomorphism $D^m f \to \{\partial^\alpha f : |\alpha| = m\}$ with some closed linear subspace of $(L_2(\Omega))^N$, where

$$N = \binom{n+m-1}{m} \tag{3.3}$$

is the number of multi-indices $\alpha = (\alpha_1,\dots,\alpha_n) \in N^n$ satisfying $|\alpha| = m$.

As for the concrete *representation formulas* of type (2.12) which may prove suitable to practical applications, it seems that the one proposed in [5] is a most natural candidate: indeed it simply follows from the classical *Taylor formula of the m-th order about the point* $a \in \bar{\Omega}$ (with integral expression of the remainder)

$$f(x) = \sum_{|\alpha|=0}^{m-1} \partial^\alpha f(a)(x-a)^\alpha/\alpha! + \int_0^1 \frac{(1-t)^{m-1}}{(m-1)!} D^m f(a+t(x-a)).(x-a)^m dt \quad , \tag{3.4}$$

which holds at every $x \in \bar{\Omega}$ for every $f \in C^m(\bar{\Omega})$, by integration over Ω with respect to the Lebesgue measure da (needless to say, other measures could be used as well); in view of the *Sobolev imbedding theorem*, the resulting *pointwise* definitions

$$Pf(x) := S^{-1} \sum_{|\alpha|=0}^{m-1} \int_\Omega \{\partial^\alpha f(a)(x-a)^\alpha/\alpha!\}da \quad , \tag{3.5a}$$

$$V(D^m f)(x) := S^{-1} \int_0^1 \frac{(1-t)^{m-1}}{(m-1)!} \{\int_\Omega D^m f(a+t(x-a)).(x-a)^m da\}dt \quad , \tag{3.5b}$$

can then be extended by continuity to $H^m(\Omega)$ iff

$$m > n/2 \tag{3.5c}$$

(strictly speaking, definition (3.5a) can *always* be extended). As readily verified, every partial derivative or order $< m$ of such functions $VD^m f$, with $f \in C^m(\bar{\Omega})$, can be similarly expressed in the integral form

$$\partial^\alpha(VD^m f)(x) = S^{-1}\int_0^1 \frac{(1-t)^{m-1-|\alpha|}}{(m-1-|\alpha|)!}$$

$$\{\int_\Omega D^{m-|\alpha|}\partial^\alpha f(a+t(x-a)).(x-a)^{m-|\alpha|} da\}dt \quad , \tag{3.6a}$$

so that it depends only on (certain) m-th partial derivatives of f or, equivalently (since $f-VD^m f \in P_{m-1}$) , of $VD^m f$ itself; this pointwise definition can be extended by continuity to $H^m(\Omega)$ iff

$$|\alpha| < m - n/2 \quad . \tag{3.6b}$$

It turns out that the corresponding *direct sum decomposition* (2.11) of $H^m(\Omega)$, which is *topological* irrespective of condition (3.5), is relatively complicated to define explicitly; it amounts indeed to the concrete definition

$$X_o := \{f \in H^m(\Omega) : \sum_{|\alpha|=0}^{m-1-|\beta|} \int_\Omega \{\partial^{\alpha+\beta} f(a)(0-a)^\alpha/\alpha!\}da = 0 \quad ,$$

$$0 \le |\beta| \le m-1\} \quad . \tag{3.7}$$

This may be the main reason why the above extension of Taylor formulas to $H^m(\Omega)$ can hardly be found in the relevant literature; an exception is [7] (see p.79), whose purpose is quite different from ours, however.

As analysed in detail in [4] (the reader is referred specially to Section 3), every expression of the type $D^m f(y).(x-a)^m$ (where $y \in R^n$) can be interpreted simply as the product of the row matrix formed with the coordinates $\partial^m f(y)/\partial y_{i_1} \dots \partial y_{i_m}$ ($i_1, \dots, i_m$ run independently from 1 to n) of the (completely symmetric) covariant tensor $D^m f(y)$ and of the column matrix formed with the coordinates of the m-fold Kronecker product $(x-a)^m$, these coordinates being taken with respect to the canonical basis of $R^{(n^m)}$ and linearly ordered in some consistent way. In view of this *matrix interpretation*, it proves relatively easy to deduce from (3.6a) the sharp appraisals:

$$|\partial^\alpha VD^m f(x)| \le \left(\frac{(m-|\alpha|)S^{-1}}{(m-|\alpha|-n/2)}\right)^{1/2} \frac{h^{m-|\alpha|}}{(m-|\alpha|)!} |f|_m, \quad \forall x \in \bar{\Omega} \quad ,$$

$$(3.8)$$

which hold, *under condition* (3.6b), for all $f \in H^m(\Omega)$. It must be emphasized that this result is *optimal*: indeed it directly follows by applying the Schwarz inequality (for n-dimensional integrals over Ω) to some trivial upper bound of $|\partial^\alpha VD^m f(x)|$, which itself can be regarded as the sharpest possible; in fact, this preliminary appraisal is obtained by first applying the Cauchy-Schwarz inequality to the scalar product under the integral sign $\int_\Omega$ in (3.6a) and substituting uniformly (with respect to a and x ranging over $\bar{\Omega}$) the upper bound h for the Euclidean norm of $x-a$, then setting $y = a+t(x-a)$ so that the original 'double' integral is transformed into an integral over the $(n+1)$-dimensional cone in $\mathcal{R}^{n+1}$ with base $\Omega \times \{0\}$ and vertex $(x,1)$, and finally integrating with respect to t for each fixed $y \in \bar{\Omega}$.

The fact mentioned above that the linear projector P of $H^m(\Omega)$ onto $\mathcal{P}_{m-1}$ defined by (3.5a) is continuous irrespective of (3.5) or, equivalently, that the related direct sum decomposition of $H^m(\Omega)$ is always topological, is known to imply (by the Rellich compactness lemma) a version of the Bramble-Hilbert lemma which asserts the *equivalence over the linear subspace* X_o *defined by* (3.7) *of the Sobolev norms* $|.|_m$ *and* $\|.\|_m$ *defined by* (3.1a,b). So, a priori, appraisals of the type

$$|VD^m f|_j \leq d(m,n,j)\, \frac{h^{m-j}}{(m-j)!}\, |f|_m \qquad (3.9a)$$

must hold for all $f \in H^m(\Omega)$ and $0 \leq j \leq m-1$, with $d(m,n,j)$ denoting some finite positive constant. Unlike appraisals (3.8), these estimates cannot be optimized at reasonable cost: indeed, for each fixed (m,n,j), the smallest possible constant d_o in (3.9a) must be such that d_o^{-2} is the lowest point of the spectrum of some differential operator, its exact determination requiring in fact the solution of a certain complicated boundary value problem. On the other hand, it proves fairly easy to deduce from (3.6a) upper bounds $d(m,n,j)$ for d_o, viz.,

$$d(m,n,j) := (m-j) \min_p \left(\frac{\int_0^1 (1-t)^{2p} \min[t^{-n},(1-t)^{-n}]dt}{2m-2j-2p-1} \right)^{1/2} \qquad (3.9b)$$

This result follows by first applying the Cauchy-Schwarz inequality under the integral sign $\int_\Omega$ in (3.6a) and substituting h for the Euclidean norm of $x-a$, then applying the Schwarz inequality to the square of the preliminary upper bound so obtained (which 'double' integral over the (n+1)-dimensional cylinder $\Omega \times (0,1)$ in $\mathcal{R}^{n+1}$ is to be interpreted as a scalar product with respect to the measure $(1-t)^{2p}dtda$ for $2p > -1$), and finally setting $y = a+t(x-a)$ in $\int_\Omega \int_\Omega |D^m f(a+t(x1a))|^2 dxda$ regarded as an iterated integral where the a-integration (resp. x-integration) is to be performed first if $t \le 1/2$ (resp. $t \ge 1/2$) so that eventually it is bounded above by the simple expression $\min[t^{-n},(1-t)^{-n}]S|f|_m^2$. Various upper bounds can be given for $d(m,n,j)$ as defined by (3.9b), in particular the following:

$$d(m,n,j) < 2(m.j)2^{n/2}/2^{m-j} \quad , \tag{3.9c}$$

$$d(m,n,j) < \left(\frac{2^{n-1}-1}{n-1}\right)^{1/2} \frac{m-j}{(2m-2j-2)^{1/2}} \quad \text{if} \quad m-j \ge 2 \quad , \tag{3.9d}$$

$$d(m,n,j) < \left(2\,\frac{2^{n-1}-1}{n-1}\right)^{1/2} \quad \text{if} \quad m-j = 1 \quad , \tag{3.9d bis}$$

and of course those resulting directly from (3.8), viz.,

$$d(m,n,j) \le \left(\frac{m-j}{m-j-n/2}\right)^{1/2} \quad \text{if} \quad m-j > n/2 \quad . \tag{3.9e}$$

4. A PRACTICAL METHOD FOR BOUNDING REMAINDERS IN LINEAR APPROXIMATION

The actual significance, as error bounding tool, of the key estimates (3.8) and (3.9) will be illustrated here by their application to the classical problem of bounding above, in terms of $|f|_2$, the Sobolev seminorm $|.|_1$ of the error made over a given triangle $\bar{\Omega} \subset \mathcal{R}^2$ when an arbitrary function $f \in H^2(\Omega)$ is interpolated at the vertices x_i $(i = 1,2,3)$ by an affine function Πf ; it is known that

$$\Pi f(x) := \sum_{i=1}^{3} f(x_i)p_i(x) \quad , \quad x \in \bar{\Omega} \quad , \tag{4.1}$$

where $p_i(x)$ is simply the i-th barycentric coordinate of x with respect to the x_i (problems of this type arise most naturally in connection with the *nodal finite element method*, where indeed piecewise polynomial interpolants over triangulated domains play a leading role). Since $\mathcal{P}_1 \subset \text{kernel } (1_X - \Pi)$, we readily get

$$\begin{aligned} |f-\Pi f|_1 &\le |VD^2 f|_1 + \sum_{i=1}^{3} |VD^2 f(x_i)||p_i|_1 \\ &\le 2^{1/2} h|f|_2 + (2S)^{-1/2} \left(\sum_{i=1}^{3} |p_i|_1\right) h^2 |f|_2 \quad , \end{aligned} \tag{4.2a}$$

by first substituting VD^2f for f in the *interpolation error* $f-\Pi f$, then applying the triangle inequality for $|.|_1$, and finally making use of appropriate key estimates of type (3.8),(3.9). At this stage, necessarily specific arguments (pertaining to 'hard analysis') are to be used to proceed farther; in the present example, it can be proved by elementary geometry that

$$\sum_{i=1}^{3} |p_i|_1 = 2S^{1/2}/\rho \quad , \tag{4.2b}$$

where ρ denotes the Euclidean diameter of the inscribed sphere of Ω , so that finally

$$|f-\Pi f|_1 \le 2^{1/2}(1+h/\rho)h|f|_2 < 2^{3/2}/(h^2/\rho)|f|_2 \quad , \quad \forall f \in H^2(\Omega) \quad . \tag{4.3}$$

Compared with the precise estimate for arbitrary triangles that can follow from Theorem 1.2 in [8], (4.3) is only 25% less sharp, which proves surprisingly good; on the other hand, for the standard rectangular triangle, the coefficient of $|f|_2$ in (4.3) reduces to $2^{3/2}(1+2^{1/2})$, which is much larger than the upper bound of 0.81 given in [8] (see Theorem 1.1) for the same theoretical error coefficient c_o. It should be stressed, however, that unlike Natterer in [8], we are not primarily interested here in finding sharp estimates of one *specific* c_o (by necessarily ingenious, specialized techniques), but rather in devising a *practical method of wide applicability for quantitative error estimation.*

In conclusion, we will describe in general terms the method of that type which comes out from the above theoretical analysis and concrete (sample) illustration. The first stage consists in selecting a *topological direct sum decomposition formula* of type (2.11) or, equivalently, a *representation formula* of type (2.12) with V being a linear right inverse of the given linear surjection $U: X \to Y$ and $P \equiv 1_X - VU$ a linear *continuous* projector of X onto $K \equiv$ kernel U (such a selection is always possible whenever K is finite-dimensional). As a general rule, it turns out that the restriction R_r to $X_o \equiv VU(X)$ of the *specific* linear mapping $R: X \to Z$ can be regarded as a finite sum

$$R_r = \sum_j S_j \tag{4.4}$$

of *standard* linear mappings $S_j: X_o \to Z$ such that each composite $S_j V: Y \to Z$ is *continuous* (typically, S_j is a certain partial derivative, possibly composed with pointwise multiplication by given functions or integration with respect to given measures). The final stage amounts to determining constants d_j such that

$$\|S_j VUf\|_Z \leq d_j \|Uf\|_Y \quad , \qquad \forall f \in X \quad , \tag{4.5}$$

either directly or by making use of suitable *key estimates*. The requested quantitative estimate

$$\|Rf\|_Z \leq (\sum_j d_j) \|Uf\|_Y \quad , \qquad \forall f \in X \quad , \tag{4.6}$$

which trivially follows is of type (2.1) rather than of type (2.14); the welcome fact that the error coefficient in (4.6) is obtained without ever trying to evaluate $\|R\|$ directly must be emphasized, the more so that assumption (H2) (unlike assumption (H1)!) need not be fulfilled here.

REFERENCES

1. Arcangéli, R. and Gout, J.L. (1976). Sur l'évaluation de l'erreur d'interpolation de Lagrange dans un ouvert de $\mathcal{R}^n$, *R.A.I.R.O. Analyse Numérique* 10, 5-27.
2. Ciarlet, P.G. and Raviart, P.A. (1972). General Lagrange and Hermite Interpolation in $\mathcal{R}^n$ with Applications to Finite Element Methods, *Arch. Rational Mech. Anal.* 46, 177-199.
3. Gout, J.L. (1976). Sur l'estimation de l'erreur d'interpolation dans $\mathcal{R}^n$, *Thèse 3e cycle*, Université de Pau.
4. Meinguet, J. (1975). Realistic Estimates for Generic Constants in Multivariate Pointwise Approximation, *in* "Topics in Numerical Analysis II" (Ed. J.J.H. Miller) 89-107, Academic Press, London, New York.
5. Meinguet, J. (1978). Structure et estimations de coefficients d'erreurs, *R.A.I.R.O. Analyse Numérique*. To appear.
6. Meinguet, J. and Descloux, J. (1977). An Operator-Theoretical Approach to Error Estimation, *Numer. Math.* 27, 307-326.
7. Morrey, C.B. (1966). Multiple Integrals in the Calculus of Variations. Springer-Verlag, Inc., New York.

8. Natterer, F. (1975). Berechenbare Fehlerschranken für die Methode der finiten Elements, *International Series of Numerical Mathematics* 28, 109-121. Birkhaüser Verlag, Basel.

APPROXIMATION OF BIVARIATE FUNCTIONS BY MEANS OF SOME BERNSTEIN-TYPE OPERATORS

D.D. Stancu

University of Cluj, Rumania

1. INTRODUCTION

It is well known that the Bernstein operators represent a class of remarkable linear positive operators which permit a simple and constructive proof for the Weierstrass approximation theorem. They provide simultaneous approximation procedures for a function and its derivatives. Perhaps the reason that they have not been widely used in practice is the poor convergence properties of their sequence. But it should be recalled that the Bernstein polynomials, corresponding to a continuous function, possess many beautiful properties such as: variation diminishing - in the sense of Schoenberg[10], shape-preserving, etc.

As W.J. Gordon and R.F. Riesenfeld have pointed out in a recent paper[7], in a large class of problems the smoothness of an approximating function is of greater importance than closeness of fit. This is illustrated with problems of computer-aided geometric design of curves and surfaces, where aesthetic reasons and the intrinsic properties of shape are essential.

Significant contributions in this field were given by P. Bézier[1], who has exploited the properties of the parametric Bernstein polynomials, used as control variables in the design of free-form curves for application such as the design of exterior surfaces of automobiles, ships and aircraft. One should mention also the important and central contributions of A.R. Forrest[3,4] in the investigation of a class of problems called by him: *computational geometry*, which is concerned with the representation in computers and manipulation by them of shape information.

We mention that the basic method of Bézier for constructing smooth curves is intimately connected with an iterated linear interpolation interpretation of the Bernstein polynomials (see G. Strang[20] and D.D. Stancu[19]), which will be extended in this paper to the case of a triangular region.

Taking into account the new applications of the Bernstein polynomials, it is important and useful to construct and to investigate new Bernstein-type polynomials in one and several variables.

2. A CLASS OF BERNSTEIN-TYPE BIVARIATE POLYNOMIALS

In a paper presented at the Symposium on Approximation Theory held at the University of Lancaster in 1969[15] we gave a probabilistic method for constructing an operator $P_m^{<\alpha>}$, depending on a real parameter α, defined - in the case of two variables - on the vector space $C(T)$ of functions f continuous on the triangle $T = \{(x,y) \mid x \geq 0,\ y \geq 0,\ x+y \leq 1\}$; it is defined by:

$$(L_m^{<\alpha>} f)(x,y) = \sum_{i=0}^{m} \sum_{j=0}^{m-i} w_{m,i,j}^{<\alpha>}(x,y)\, f\left(\frac{i}{m}, \frac{j}{m}\right) , \qquad (2.1)$$

with

$$w_{m,i,j}^{<\alpha>}(x,y) = \binom{m}{i} \binom{m-i}{j} \frac{x^{(i,-\alpha)}\, y^{(j,-\alpha)}\, (1-x-y)^{(m-i-j,-\alpha)}}{1^{(m,-\alpha)}} ,$$

where by $u^{(n,h)}$ one denotes the factorial power of order n and increment h of u.

That method involves the Markov-Pólya probability distribution.

For $\alpha = 0$ it reduces to the Bernstein polynomial of overall degree n for the simplex T:

$$(B_m f)(x,y) = \sum_{i=0}^{m} \sum_{j=0}^{m-i} \binom{m}{i} \binom{m-i}{j} x^i y^i (1-x-y)^{m-i-j} f(\frac{i}{m}, \frac{j}{m}) \quad . \tag{2.2}$$

It should be noticed that $L_m^{<\alpha>}$ can be represented[14] by means of B_m under the following form

$$(L_m^{<\alpha>} f)(x,y) = \frac{1}{B(\frac{x}{\alpha}, \frac{y}{\alpha}, \frac{1-x-y}{\alpha})} \iint_T u^{\frac{x}{\alpha}-1} v^{\frac{y}{\alpha}-1} (1-u-v)^{\frac{1-x-y}{\alpha}-1} (B_m f)(u,v) \, dudv \quad ,$$

where $\alpha > 0$, $x > 0$, $y > 0$, $x+y < 1$, and $B(a,b,c)$ denotes the Dirichlet double integral, that is:

$$B(a,b,c) = \iint_T u^{a-1} v^{b-1} (1-u-v)^{c-1} \, dudv \quad .$$

Hence $L_m^{<\alpha>}$ can be looked upon as a certain average of the Bernstein operator B_m.

In her doctoral thesis[8], Rosina Günttner generalized our operator $L_m^{<\alpha>}$ in the sense in which H. Brass[2] has generalized the Bernstein univariate operators.

In the next section we shall see that the operator $L_m^{<\alpha>}$ represents a special case of a general class of linear positive operators of Bernstein-type.

3. AN INTERPOLATING METHOD FOR CONSTRUCTING BERNSTEIN-TYPE BIVARIATE POLYNOMIALS

Now we give an interpolating method for obtaining Bernstein polynomials in two variables. The extension to more than two variables is immediate.

We shall start from the Steffensen interpolating formula[21,13] corresponding to a function f, defined on a polynomial domain Ω, and the array of nodes $(t_i, z_j) \in \Omega$, where $i = 0(1)m$, $j = 0(1)n_i$ and $n_o \geq n_1 \geq \ldots \geq n_m$:

$$f(t,z) = (Sf)(t,z) + (Rf)(t,z) \quad , \tag{3.1}$$

where

$$(Sf)(t,z) = \sum_{i=0}^{m} \sum_{j=0}^{n_i} (t-t_o) \ldots (t-t_{i-1})(z-z_o) \ldots (z-z_{j-1}) \left[\begin{matrix} t_o, t_1, \ldots, t_i \\ z_o, z_1, \ldots, z_j \end{matrix} ; f \right] ,$$

and

$$(Rf)(t,z) = (t-t_o)(t-t_1)\ldots(t-t_m)[t, t_o, t_1, \ldots, t_m; f] +$$

$$+ \sum_{i=0}^{m} (t-t_o)\ldots(t-t_{i-1})(z-z_o)\ldots(z-z_{n_i}) \left[\begin{matrix} t_o, t_1, \ldots, t_i \\ z, z_o, \ldots, z_{n_i} \end{matrix} ; f \right] ,$$

the brackets representing the symbol for divided differences.

We shall choose as function f a polynomial $\phi^{<\alpha,\beta>}$, whose coefficients might depend on two real parameters α and β, such that the remainder $R\phi^{<\alpha,\beta>}$ vanishes.

Assuming that the coordinates of the nodes are equally spaced: $t_i = x+i\alpha$, $z_j = y+j\beta$ $(i=0(1)m,\ j=0(1)n_i)$, where (x,y) is any fixed point of Ω, we obtain the following representation

$$\phi^{<\alpha,\beta>}(t,z) = \sum_{i=0}^{m} \sum_{j=0}^{n_i} \frac{(t-x)^{(i,\alpha)}(z-y)^{(j,\beta)}}{i!\, j!\, \alpha^i\, \beta^j} \Delta_{\alpha,\beta}^{i,j}\, \phi^{<\alpha,\beta>}(x,y),$$

where

$$\Delta_{\alpha,\beta}^{i,j}\, \phi^{<\alpha,\beta>}(x,y) = \Delta_{1,\alpha}^{i}(\Delta_{2,\beta}^{j}\phi^{<\alpha,\beta>}(x,y)) =$$

$$\sum_{\nu=0}^{i} \sum_{\mu=0}^{j} (-1)^{\nu+\mu}\binom{i}{\nu}\binom{j}{\mu}\phi^{<\alpha,\beta>}(x+(i-\nu)\alpha, y+(j-\mu)\beta) \quad .$$

If we choose $t = z = 0$ and take into account that

$$(-x)^{(i,\alpha)} = (-1)^i x^{(i,-\alpha)}, (-y)^{(j,\beta)} = (-1)^j y^{(j,-\beta)} \quad ,$$

we can write

$$\phi^{<\alpha,\beta>}(0,0) = \sum_{i=0}^{m} \sum_{j=0}^{n_i} (-1)^{i+j} \frac{x^{(i,-\alpha)} y^{(j,-\beta)}}{i!\, j!\, \alpha^i\, \beta^j} \Delta_{\alpha,\beta}^{i,j}\, \phi^{<\alpha,\beta>}(x,y).$$

Let us introduce the Nörlund difference quotients $D_{1,\alpha}, D_{2,\beta}$, defined by:

$$D_{1,\alpha}g(x,y) = \frac{1}{\alpha}[g(x+\alpha,y) - g(x,y)] \quad ,$$

$$D_{2,\beta}g(x,y) = \frac{1}{\beta}[g(x,y+\beta) - g(x,y)] \quad .$$

Positive integral powers of these operators are defined by iteration.

Using the notation $D_{1,\alpha}^{i} D_{2,\beta}^{j} = D_{\alpha,\beta}^{i,j}$, and assuming that $\phi^{<\alpha,\beta>}(0,0) \neq 0$, we can write the following identity

$$\frac{1}{\phi^{<\alpha,\beta>}(0,0)} \sum_{i=0}^{m} \sum_{j=0}^{n_i} (-1)^{i+j} \frac{x^{(i,-\alpha)} y^{(j,-\beta)}}{i!\, j!} D_{\alpha,\beta}^{i,j}\, \phi^{<\alpha,\beta>}(x,y) = 1 \; .$$

We now associate to each function f, defined on Ω, a linear operator $L^{<\alpha,\beta>}$ depending on the parameters α and β, given by

$$(L^{<\alpha,\beta>}f)(x,y) = \frac{1}{\phi^{<\alpha,\beta>}(0,0)} \sum_{i=0}^{m} \sum_{j=0}^{n_i} (-1)^{i+j} \frac{x^{(i,-\alpha)} y^{(j,-\beta)}}{i!\, j!} D_{\alpha,\beta}^{i,j} \phi^{<\alpha,\beta>}(x,y) \cdot f(x_i,y_j) ,$$

where $(x_i,y_j) \in \Omega$ $(i=0(1)m,\ j=0(1)n_i)$.

It is easy to see that: $(L^{<\alpha,\beta>}f)(0,0) = f(x_o,y_o)$.

If we make the further assumptions that $\alpha \geq 0$, $\beta \geq 0$ and $(-1)^{i+j} D_{\alpha,\beta}^{i,j} \phi^{<\alpha,\beta>}(x,y) \geq 0$ for $i=0(1)m$, $j=0(1)n_i$ and $(x,y) \in \Omega$ then $(L^{<\alpha,\beta>})$ represents a family of positive linear operators.

We mention two remarkable special cases of these operators:

(i) $\Omega = S = [0,1] \times [0,1], n_i = n,\ \phi^{<\alpha,\beta>}(x,y) = \phi^{<\alpha>}(x) \cdot \Psi_n^{<\beta>}(y)$;

(ii) $\Omega = T = \{(x,y) \mid x \geq 0,\ y \geq 0,\ x+y \leq 1\},\ \beta = \alpha,\ n_i = m-i,$

$$\phi^{<\alpha,\beta>}(x,y) = \phi_m^{<\alpha>}(x,y) .$$

In the first case we obtain a double sequence of operators defined by

$$(L_{m,n}^{<\alpha,\beta>}f)(x,y) = \frac{1}{\phi_m^{<\alpha>}(0)\ \Psi_n^{<\beta>}(0)} \sum_{i=0}^{m} \sum_{j=0}^{n} (-1)^{i+j} \frac{x^{(i,-\alpha)} y^{(j,-\beta)}}{i!\, j!} (D_{1,\alpha}^{i}\ \phi_m^{<\alpha>}(x)) (D_{2,\beta}^{j}\ \Psi_n^{<\beta>}(y)) f(x_{m,i},y_{n,j}) , \quad (3.2)$$

where (x,y) and $(x_{m,i},y_{n,j})$ belong to the square S.

In the second case we get the simple sequence of operators defined by:

$$(L_m^{<\alpha>} f)(x,y) =$$

$$\frac{1}{\phi_m^{<\alpha>}(0)} \sum_{i=0}^{m} \sum_{j=0}^{m-i} (-1)^{1+j} \frac{x^{(i,-\alpha)} y^{(j,-\alpha)}}{i!\ j!} (D_{\alpha,\alpha}^{i,j}\ \phi_m^{<\alpha>}(x,y))$$

$$f(x_{m,i},y_{m,j}) \quad , \qquad (3.3)$$

where (x,y) and $(x_{m,i},y_{m,j})$ belong to the triangle T.

We now state and prove

THEOREM 3.1

In the case (i) for the nodes $(\frac{i}{m}, \frac{j}{n})$ and for any $(x,y) \in S$, the approximating polynomial can be represented in the following form:

$$(L_{m,n}^{<\alpha,\beta>} f)(x,y) -$$

$$\frac{1}{\phi_m^{<\alpha>}(0)\Psi_n^{<\beta>}(0)} \sum_{i=0}^{m} \sum_{j=0}^{n} (-1)^{i+j} \frac{x^{(i,-\alpha)} y^{(j,-\beta)}}{i!\ j!}$$

$$(D_{1,\alpha}^{i}\phi_m^{<\alpha>}(0))(D_{2,-\beta}^{j}\Psi_n^{<\beta>}(0))\ \Delta_{\frac{1}{m},\frac{1}{n}}^{i,j} f(0,0) \quad . \qquad (3.4)$$

In the case (ii) for the nodes $(\frac{i}{m}, \frac{j}{m})$ and for any $(x,y) \in T$, the approximating polynomial can be written under the form:

$$(L_m^{<\alpha>} f)(x,y) =$$

$$\frac{1}{\phi^{<\alpha>}(0,0)} \sum_{i=0}^{m} \sum_{j=0}^{m-i} (-1)^{i+j} \frac{x^{(i,-\alpha)} y^{(j,-\alpha)}}{i!\, j!} (D^{i,j}_{-\alpha,-\alpha} \phi_m^{<\alpha>}(0,0)) \cdot$$

$$\cdot \Delta^{i,j}_{\frac{1}{m},\frac{1}{m}} f(0,0) \quad . \tag{3.5}$$

Proof: Because $L^{<\alpha,\beta>}_{m,n} f$ is a polynomial of degree (m,n), if we use as coordinates of the notes: $t_i = -i\alpha$ $(i=0(1)m)$, $z_j = -j\beta$ $(j=0(1)n)$, then by taking $t = x$ and $z = y$, the Steffensen interpolation formula (3.1) becomes the Newton bivariate interpolation formula and we can write

$$(L^{<\alpha,\beta>}_{m,n} f)(x,y) = \sum_{i=0}^{m} \sum_{j=0}^{n} x^{(i,-\alpha)} y^{(j,-\beta)} \left[\begin{matrix} 0,-\alpha,\ldots,-i\alpha \\ 0,-\beta,\ldots,-j\beta \end{matrix} ; (L^{<\alpha,\beta>}_{m,n} f)(u,v)\right] ,$$

because the corresponding remainder vanishes.

Taking into account that

$$\left[\begin{matrix} 0,-\alpha,\ldots,-i\alpha \\ 0,-\beta,\ldots,-j\beta \end{matrix} ; (L^{<\alpha,\beta>}_{m,n} f)(u,v)\right] = \frac{1}{i!j!} D^{i,j}_{-\alpha,-\beta} (L^{<\alpha,\beta>}_{m,n} f)(0,0) ,$$

we obtain:

$$(L^{<\alpha,\beta>}_{m,n} f)(x,y) =$$

$$\sum_{i=0}^{m} \sum_{j=0}^{n} \frac{x^{(i,-\alpha)} y^{(j,-\beta)}}{i!\, j!} D^{i,j}_{-\alpha,-\beta} (L^{<\alpha,\beta>}_{m,n} f)(0,0) \quad . \tag{3.6}$$

According to a lemma stated and proved in our previous paper[16], we have

$$(-1)^{i+j}\phi_m^{<\alpha>}(0)\Psi_n^{<\beta>}(0)D_{-\alpha,-\beta}^{i,j}(L_{m,n}^{<\alpha,\beta>}f)(x,y) =$$

$$\sum_{\nu=0}^{m-i}\sum_{\mu=0}^{n-j}(-1)^{\nu+\mu}\frac{x^{(\nu,-\alpha)}y^{(\mu,-\beta)}}{\nu!\ \mu!}\times$$

$$D_{1,\alpha}^{\nu}(D_{1,-\alpha}^{i}\phi_m^{<\alpha>}(x))D_{2,\beta}^{\mu}(D_{2,-\beta}^{j}\Psi_n^{<\beta>}(y))\Delta_{\frac{1}{m},\frac{1}{n}}^{i,j}f(0,0) ,$$

where $0 \leq i \leq m$, $0 \leq j \leq n$.

Putting here $x = y = 0$, we find that

$$D_{-\alpha,-\beta}^{i,j}(L_{m,n}^{<\alpha,\beta>}f)(0,0) =$$

$$\frac{1}{\phi_m^{<\alpha>}(0)\Psi_n^{<\beta>}(0)}(D_{1,-\alpha}^{i}\phi_m^{<\alpha>}(0))(D_{2,-\beta}^{j}\Psi_n^{<\beta>}(0))\Delta_{\frac{1}{m},\frac{1}{n}}^{i,j}f(0,0) .$$

Inserting this result in (3.6) we obtain the desired formula (3.4). In the case (ii), according the formula (3.3), we have:

$$(L_m^{<\alpha>}f)(x,y) = \sum_{i=0}^{m}\sum_{j=0}^{m-i}x^{(i,-\alpha)}y^{(j,-\alpha)}\begin{bmatrix}0,-\alpha,\ldots,-i\alpha\\0,-\alpha,\ldots,-j\alpha\end{bmatrix}; L_m^{<\alpha>}f\Bigg] =$$

$$\sum_{i=0}^{m}\sum_{j=0}^{m-i}(-1)^{i+j}\frac{x^{(i,-\alpha)}y^{(j,-\beta)}}{i!\ j!}D_{-\alpha,-\alpha}^{i,j}(L_m^{<\alpha>}f)(0,0) . \quad (3.7)$$

By making use of the same lemma from [16], we obtain

$$D_{-\alpha,-\alpha}^{i,j}(L_m^{<\alpha>}f)(x,y) -$$

$$\frac{(-1)^{i+j}}{\phi_m^{<\alpha>}(0,0)}\sum_{\nu=0}^{m-i}\sum_{\mu=0}^{m-\nu-j}(-1)^{\nu+\mu}\frac{x^{(\nu,-\alpha)}y^{(\mu,-\alpha)}}{\nu!\ \mu!}$$

$$D_{\alpha,\alpha}^{\nu,\mu}(D_{-\alpha,-\alpha}^{i,j}\phi_m^{<\alpha>}(x,y))\ \Delta_{\frac{1}{m},\frac{1}{m}}^{i,j}f(0,0) ,$$

which for $x = y = 0$ reduces to

$$D^{i,j}_{-\alpha,-\alpha}(L^{<\alpha>}_m f)(0,0) =$$

$$\frac{(-1)^{i+j}}{\phi^{<\alpha>}_m(0,0)} D^{i,j}_{-\alpha,-\alpha} \phi^{<\alpha>}_m(0,0) \quad \Delta^{i,j}_{\frac{1}{m},\frac{1}{m}} f(0,0) \quad .$$

The insertion of it into (3.7) leads to the required formula (3.5).

We are now ready to state and prove the following convergence theorem.

THEOREM 3.2

i) Assume that $0 \leq \alpha = \alpha(m) \to 0$ as $m \to \infty$ and $0 \leq \beta = \beta(n) \to 0$ as $n \to \infty$. If the nodes are $(\frac{i}{m},\frac{j}{n})$ and for any natural numbers m and n the polynomials $\phi^{<\alpha>}_m$ and $\psi^{<\beta>}_n$ satisfy the conditions:

$$\phi^{<\alpha>}_m(0) > 0 \ , \quad \psi^{<\beta>}_n(0) > 0 \quad ,$$

$$(-1)^i D^i_{1,\alpha}\phi^{<\alpha>}_m(x) \geq 0, \ (-1)^j D^j_{2,\beta}\psi^{<\beta>}_n(y) \geq 0, \ \forall \ (x,y) \in S,$$

$$\lim_{m\to\infty} \frac{D_{1,-\alpha}\phi^{<\alpha>}_m(0)}{m\phi^{<\alpha>}_m(0)} = -1, \quad \lim_{n\to\infty} \frac{D_{2,-\beta}\psi^{<\beta>}_n(0)}{n\psi^{<\beta>}_n(0)} = -1,$$

$$\lim_{m\to\infty} \frac{D^2_{1,-\alpha}\phi^{<\alpha>}_m(0)}{m^2\phi^{<\alpha>}_m(0)} = 1, \quad \lim_{n\to\infty} \frac{D^2_{2,-\beta}\psi^{<\beta>}_n(0)}{n^2\psi^{<\beta>}_n(0)} = 1 \ ,$$

then for any $f \in C(S)$ the sequence $(L^{<\alpha,\beta>}_{m,n} f)$ converges uniformly to f on S.

ii) Let $0 \leq \alpha = \alpha(m) \to 0$ as $m \to \infty$ and consider the nodes $(\frac{i}{m},\frac{j}{m})$ $(0 \leq i + j \leq m)$. If for any natural number m

the polynomials $\phi_m^{<\alpha>}$ satisfy on T the conditions:

$$\phi_m^{<\alpha>}(0,0) > 0, \quad (-1)^{i+j} D^{i,j}_{-\alpha,-\alpha} \phi_m^{<\alpha>}(x,y) \geq 0 ,$$

$$\lim_{m\to\infty} \frac{D_{1,-\alpha}\phi_m^{<\alpha>}(0,0)}{m\phi_m^{<\alpha>}(0,0)} = \lim_{m\to\infty} \frac{D_{2,-\alpha}\phi_m^{<\alpha>}(0,0)}{m\phi_m^{<\alpha>}(0,0)} = -1 ,$$

$$\lim_{m\to\infty} \frac{D^2_{1,-\alpha}\phi_m^{<\alpha>}(0,0)}{m^2\phi_m^{<\alpha>}(0,0)} = \lim_{m\to\infty} \frac{D^2_{2,-\alpha}\phi_m^{<\alpha>}(0,0)}{m^2\phi_m^{<\alpha>}(0,0)} = 1 ,$$

then for any $f \in C(T)$ the sequence $(L_m^{<\alpha>} f)$ converges uniformly to f on T.

Proof: We shall make use of the well-known Bohman-Korovkin theorem for two variables (see, e.g. [5,9]). We first note that for the test functions $e_{i,j}$ $(0 \leq i+j \leq 2)$, where $e_{i,j}(x,y) = x^i y^j$, formula (3.4) permits us to find immediately that

$$(L_m^{<\alpha,\beta>} e_{0,0})(x,y) = 1, \quad (L_{m,n}^{<\alpha,\beta>} e_{1,0}) = -x \frac{D_{1,-\alpha}\phi_m^{<\alpha>}(0)}{m\phi_m^{<\alpha>}(0)} ,$$

$$(L_{m,n}^{<\alpha,\beta>} e_{0,1})(x,y) = -y \frac{D_{2,-\beta}\Psi_n^{<\beta>}(0)}{n\Psi_n^{<\beta>}(0)} ,$$

$$(L_{m,n}^{<\alpha,\beta>} e_{1,1})(x,y) = \frac{D_{1,-\alpha}\phi_m^{<\alpha>}(0)}{m\phi_m^{<\alpha>}(0)} \frac{D_{2,-\beta}\Psi_n^{<\beta>}(0)}{n\Psi_n^{<\beta>}(0)} xy ,$$

$$(L_{m,n}^{<\alpha,\beta>} e_{2,0})(x,y) = x(x+\alpha) \frac{D^2_{1,-\alpha}\phi_m^{<\alpha>}(0)}{m^2\phi_m^{<\alpha>}(0)} - \frac{D_{1,-\alpha}\Psi_n^{<\alpha>}(0)}{m^2\Psi_n^{<\alpha>}(0)} x ,$$

$$(L_{m,n}^{<\alpha,\beta>} e_{0,2})(x,y) = y(y+\beta) \frac{D^2_{2,-\beta}\Psi_n^{<\beta>}(0)}{n^2\Psi_n^{<\beta>}(0)} - \frac{D_{2,-\beta}\Psi_n^{<\beta>}(0)}{n^2\Psi_n^{<\beta>}(0)} y .$$

Now referring to our hypotheses, we see at once that we have uniformly on S :

$$\lim_{m,n\to\infty} L_{m,n}^{<\alpha,\beta>} e_{i,j} = e_{i,j} \qquad (0 \le i+j \le 2) \quad .$$

According to the theorem of Bohman-Korovkin we can state that the sequence $(L_{m,n}^{<\alpha,\beta>} f)$ converges uniformly to f on S is $f \in C(S)$.

In the case of the operators defined at (3.3) formula (3.5) permits us to find that for the test functions $e_{i,j}$ $(0 \le i+j \le 2)$ we obtain similar expressions, from which it is easily seen that we have uniformly on the triangle T :

$$\lim_{m\to\infty} L_m^{<\alpha>} e_{i,j} = e_{i,j} \qquad (0 \le i+j \le 2) \quad .$$

Invoking again the Bohman-Korovkin theorem, there follows that if $f \in C(T)$ then $L_m^{<\alpha>} f \to f$ uniformly on T .

Examples

a) If we assume that

$$\phi_m^{<\alpha>}(x) = (1-x)^{(m,-\alpha)} \quad , \quad \psi_n^{<\beta>}(y) = (1-y)^{(n,-\beta)} \quad ,$$

and that we have the nodes $(\frac{i}{m},\frac{j}{n})$, then according to (3.2) we obtain the Bernstein-type operator defined by:

$$(L_{m,n}^{<\alpha,\beta>} f)(x,y) = \sum_{i=0}^{m} \sum_{j=0}^{n} w_{m,i}^{<\alpha>}(x) w_{n,j}^{<\beta>}(y) f(\frac{i}{m},\frac{j}{n}) \quad , \tag{3.8}$$

where we use the notation

$$w_{p,k}^{<\gamma>}(t) = \binom{p}{k} \frac{t^{(k,-\gamma)}(1-t)^{(p-k,-\gamma)}}{1^{(p,-\gamma)}} \quad . \tag{3.9}$$

b) If we select $\phi_m^{<\alpha>}(x,y) = (1-x-y)^{(m,-\alpha)}$ and choose the nodes $(\frac{i}{m},\frac{j}{n})$ $(0 \le i+j \le m)$, then formula (3.3) leads us to the operators defined at (2.1).

It is easy to see that the convergence conditions stated in Theorem 3.2 are satisfied in these two special cases.

4. TENSOR PRODUCT AND BOOLEAN SUM OPERATORS

By using the algebraic approach to the approximation of multivariate functions described by W.J. Gordon[6], we can interpret the operator defined at (3.2) as the *tensor product* (cross product) of the univariate linear operators $L_m^{<\alpha>}$ and $L_n^{<\beta>}$ investigated in our paper[16]: $L_{m,n}^{<\alpha,\beta>} = L_m^{<\alpha>} \otimes L_n^{<\beta>}$. It should be noted that if f is continuous on S then $L_m^{<\alpha>}$ and $L_n^{<\beta>}$ commute under multiplication.

One can also extend our operators by forming the *Boolean sum* operator namely

$$L_m^{<\alpha>} \oplus L_n^{<\beta>} = L_m^{<\alpha>} + L_n^{<\beta>} - L_m^{<\alpha>} \otimes L_n^{<\beta>} .$$

In order to be more concrete, let us consider the case of the operators defined by

$$(L_p^{<\gamma>} g)(t) = \sum_{k=0}^{p} w_{p,k}^{<\gamma>}(t) g(\frac{k}{p}) ,$$

with the notation (3.9).

The remainders associated with the univariate operators are defined by: $R_m^{<\alpha>} = I - L_m^{<\alpha>}$, $R_m^{<\beta>} = I - L_n^{<\beta>}$, where I is the identity operator.

By considering the 'minimal' and the 'maximal' decompositions of the identity operator:

$$I = L_m^{<\alpha>} \otimes L_n^{<\beta>} + (R_m^{<\alpha>} \oplus R_n^{<\beta>}), \quad I = L_m^{<\alpha>} \oplus L_n^{<\beta>} + (R_m^{<\alpha>} \otimes R_n^{<\beta>}),$$

and by using the expressions given in our paper[16] for the remainder in the univariate case, we can state:

THEOREM 4.1

Let $f \in C^{2,2}(S)$ and consider the linear approximation formulae

$$f = L_{m,n}^{<\alpha,\beta>} f + (R_m^{<\alpha>} \oplus R_n^{<\beta>}) f \ ,$$
$$f = L_m^{<\alpha>} \oplus L_n^{<\beta>} f + (R_m^{<\alpha>} \otimes R_n^{<\beta>}) f \ . \tag{4.2}$$

On any point $(x,y) \in S$ the remainders can be respectively under the following forms:

$$(R_m^{<\alpha>} \oplus R_n^{<\beta>})(f)(x,y) =$$
$$- \frac{x(1-x)}{2m} \frac{1+m\alpha}{1+\alpha} f^{(2,0)}(\xi,y) - \frac{y(1-y)}{2n} \frac{1+n\beta}{1+\beta} f^{(0,2)}(x,\eta)$$
$$- \frac{x(1-x)y(1-y)}{4mn} \frac{(1+m\alpha)(1+n\beta)}{(1+\alpha)(1+\beta)} f^{(2,2)}(\xi,\eta) \ ,$$

$$(R_m^{<\alpha>} \otimes R_n^{<\beta>})(f)(x,y) =$$
$$\frac{x(1-x)y(1-y)}{4mn} \frac{(1+m\alpha)(1+n\beta)}{(1+\alpha)(1+\beta)} f^{(2,2)}(\xi,\eta) \ ,$$

where $\xi,\eta \in (0,1)$.

It is easy to see that the blended operator reproduces the function f along the boundary of the unit square S , i.e. $(L_m^{<\alpha>} \oplus L_n^{<\beta>})(f) = f$ on ∂S , while the tensor product operator reproduces the values of f only at the four corners of S .

By using the decompositions (4.1) and the results established in [18] in the univariate case, one can also give integral representations for the remainders of the approximation formulae (4.2), as well as evaluations for the orders of approximations of the function f, continuous on the square S.

5. AN INTERPOLATING METHOD FOR CONSTRUCTING THE BERNSTEIN POLYNOMIAL FOR A TRIANGULAR REGION

Let us denote by $L(g;a,b;c,d;x,y)$ the interpolating polynomial of degree one, in two variables x and y, corresponding to a function g and to the three nodes: $A(a,c)$, $B(b,c)$, $D(a,d)$. It is easily checked that

$$L(g;a,b;c,d;x,y) =$$

$$\frac{x-a}{b-a}\, g(b,c) + \left(1 - \frac{x-a}{b-a} - \frac{y-c}{d-c}\right) g(a,c) + \frac{y-c}{d-c}\, g(a,d) \quad . \tag{5.1}$$

Consider a function f defined on the standard triangle T with vertices $(0,0)$, $(1,0)$, $(0,1)$, and the triangular array of points Γ_m : $\left(\frac{i}{m},\frac{j}{m}\right)$ $(0 \le i+j \le m)$.

First we shall construct the linear interpolation polynomial $\phi^f_{1,m}$, corresponding to the function f and the points of Γ_m . According to (5.1) we have

$$\phi^f_{1,m}(x,y) - L\left(f; \frac{i}{m},\frac{i+1}{m}; \frac{j}{m},\frac{j+1}{m}; x,y\right) = (mx-i) f\left(\frac{i+1}{m},\frac{j}{m}\right)$$

$$+ (1-mx-my+i+j) f\left(\frac{i}{m},\frac{j}{m}\right) + (my-j) f\left(\frac{i}{m},\frac{j+1}{m}\right) \quad , \tag{5.2}$$

where $0 \le i+j \le m-1$ and (x,y) belongs to the triangle $T^{(1)}_{i,j}$ having the vertices: $\left(\frac{i}{m},\frac{j}{m}\right)$, $\left(\frac{i+1}{m},\frac{j}{m}\right)$, $\left(\frac{i}{m},\frac{j+1}{m}\right)$.

Denoting by t and z any fixed points of the interval $[0,1]$, let us consider further the following system of nodes:

$(\frac{i+t}{m},\frac{j+z}{m})$ $(0 \le i+j \le m-1)$, which, evidently, belong to T .

By using again the interpolation formula (5.1), for the function $\phi^f_{1,m}$ and the nodes

$$(\frac{i+t}{m},\frac{j+z}{m}),(\frac{i+1+t}{m},\frac{j+z}{m}),(\frac{i+t}{m},\frac{j+1+z}{m}) \quad , \tag{5.3}$$

where $0 \le i+j \le m-1$, we have

$$\phi^f_{2,m}(x,y) = \phi^f_{2,m}(x,y;t,z) = L(\phi^f_{1,m};\ \frac{i+t}{m},\frac{i+1+t}{m};\ \frac{j+z}{m},\frac{j+1+z}{m};\ x,y)$$

$$= (mx-i-t)\phi^f_{1,m}(\frac{i+1+t}{m},\frac{j+z}{m}) + (1-mx-my+i+t+j+z)\phi^f_{1,m}(\frac{i+t}{m},\frac{j+z}{m})$$

$$+ (my-j-z)\phi^f_{1,m}(\frac{i+t}{m},\frac{j+1+z}{m}), \tag{5.4}$$

with $(x,y) \in T^{(2)}_{i,j}$ $(0 \le i+j) \le m-2)$, where $T^{(2)}_{i,j}$ is the triangle with the vertices (5.3).

We next use the array of nodes

$$(\frac{i+2t}{m},\frac{j+2z}{m}) \quad (0 \le i+j \le m-2)$$

and the function $\phi^f_{2,m}$ for constructing

$$\phi^f_{3,m}(x,y) = L(\phi^f_{2,m};\ \frac{i+2t}{m},\frac{i+1+2t}{m};\ \frac{j+2z}{m},\frac{j+1+2z}{m};\ x,y)$$

$$= (mx-i-2t)\phi^f_{2,m}(\frac{i+1+2t}{m},\frac{j+2z}{m}) + (1-mx-my+i+2t+j+2z)$$

$$\phi^f_{2,m}(\frac{i+2t}{m},\frac{j+2z}{m}) + (my-j-2z)\phi^f_{2,m}(\frac{i+2t}{m},\frac{j+1+2z}{m}) \quad .$$

In general, we use the array of nodes

$$(\frac{i+(r-1)t}{m},\frac{j+(r-1)z}{m}) \quad (1 \le r \le m,\ 0 \le i+j \le m-r)$$

and we obtain:

$$\phi^f_{r,m}(x,y) = \phi^f_{r,m}(x,y;t,z)$$

$$= L(\phi^f_{r-1,m};\ \frac{i+(r-1)t}{m},\frac{i+1+(r-1)t}{m};\ \frac{j+(r-1)z}{m},\frac{j+1+(r-1)z}{m};\ x,y)$$

$$+ (mx-i-(r-1)t)\phi^f_{r-1,m}(\frac{i+1+(r-1)t}{m},\frac{j+(r-1)z}{m})$$

$$+ (1-mx-my+i+(r-1)t+j+(r-1)z)\phi^f_{r-1,m}(\frac{i+(r-1)t}{m},\frac{j+(r-1)z}{m})$$

$$+ (m-y-j-(r-1)z)\phi^f_{r-1,m}(\frac{i+(r-1)t}{m},\frac{j+1+(r-1)z}{m}) \quad ,$$

the point (x,y) belonging to the triangle $T^{(r)}_{i,j}$ with vertices

$$(\frac{i+(r-1)t}{m},\frac{j+(r-1)z}{m}),(\frac{i+1+(r-1)t}{m},\frac{j+(r-1)z}{m}),(\frac{i+(r-1)t}{m},\frac{j+1+(r-1)z}{m}).$$

According to (5.2) and (5.4) we have

$$\phi^f_{1,m}(\frac{i+t}{m},\frac{j+z}{m}) = tf(\frac{i+1}{m},\frac{j}{m}) + (1-t-z)f(\frac{i}{m},\frac{j}{m}) + zf(\frac{i}{m},\frac{j+1}{m}) \quad ,$$

$$\phi^f_{2,m}(\frac{i+2t}{m},\frac{j+2z}{m}) = t^2f(\frac{i+2}{m},\frac{j}{m}) + 2t(1-t-z)f(\frac{i+1}{m},\frac{j}{m})$$

$$+ (1-t-z)^2f(\frac{i}{m},\frac{j}{m}) + 2z(1-t-z)f(\frac{i}{m},\frac{j+1}{m})$$

$$+ 2tzf(\frac{i+1}{m},\frac{j+1}{m}) + z^2f(\frac{i}{m},\frac{j+2}{m})$$

$$= \sum_{\nu=0}^{2}\sum_{\mu=0}^{2-\nu}\binom{2}{\nu}\binom{2-\nu}{\mu}t^\nu z^\mu(1-t-z)^{2-\nu-\mu}\ f(\frac{i+\nu}{\mu},\frac{j+\nu}{m}) \quad .$$

It can be proved by induction that for any $r = 1(1)m$ we have

$$\phi^f_{r,m}(\frac{i+rt}{m},\frac{j+rz}{m};\ t,z) =$$

$$\sum_{\nu=0}^{r}\sum_{\mu=0}^{r-\nu}\binom{r}{\nu}\binom{r-\nu}{\mu}t^\nu z^\mu(1-t-z)^{r-\nu-\mu}\ f(\frac{i+\nu}{m},\frac{j+\mu}{m})\quad , \tag{5.5}$$

where $0 \le i+j \le m-r$.

We see at once that if we take $r = m$ then we have $i = j = 0$ and formula (5.5) leads us to the relation: $\phi^f_{m,m}(t,z;t,z) = (B_m f)(t,z)$, that is to the Bernstein polynomial in two variables, defined at (2.2).

Note that it is also possible to start from the Newton linear interpolation polynomial

$$L(f;a,b;c,d;x,y) =$$

$$f(a,c) + \frac{x-a}{h}\Delta^{1,0}_{h,0}\ f(a,c) + \frac{y-c}{k}\Delta^{0,1}_{0,k}\ f(a,c)\quad ,$$

where $h = b-a$ and $k = d-c$. In this case one finds similarly the following representation, in terms of finite differences, of the Bernstein polynomial for a triangle:

$$(B_m f)(x,y) = \sum_{i=0}^{m}\sum_{j=0}^{m-i}\binom{m}{i}\binom{m-i}{j}x^i y^j\ \Delta^{i,j}_{\frac{1}{m},\frac{1}{m}}\ f(0,0)\quad .$$

These methods for obtaining Bernstein-type polynomials, by starting from a linear interpolation formula, can be useful for extending the Bézier method in order to construct smooth curves and surfaces.

Finally, we mention that in our earlier paper[12] we gave a more general method for constructing Bernstein polynomials in two variables. We make the remark that by using that method one can extend also to a triangular domain the Schoenberg spline approximation formula[11] generalizing the ordinary Bernstein polynomials.

ACKNOWLEDGEMENT

The author wishes to express his gratitude to the Organizing Committee for according to him a financial support which enabled him to attend this symposium.

REFERENCES

1. Bézier, P. (1972). "Numerical Control - Mathematics and Applications" (translated by A.R. Forrest), Wiley.

2. Brass, H. (1971). Eine Verallgemeinerung de Bernsteinschen Operatoren, *Abhandl. Math. Sem. Hamburg* 36, 111-122.

3. Forrest, A.R. (1971). Computational Geometry, *Proc. Roy. Soc. London* A.321, 187-195.

4. Forrest, A.R. (1972). Interactive Interpolation and Approximation by Bézier Polynomials, *Computer J.* 15, 71-79.

5. Freud, G. (1964). Über positive lineare Approximationsfolgen von stetigen reellen Funktionen auf kompakten Mengen, *in* "On Approximation Theory", 233-238 (Eds. P.L. Butzer and J. Korevaar) Birkhäuser Verlag.

6. Gordon, W.J. (1969). Distributive Lattices and the Approximation of Multivariate Functions, *in* "Approximation with Special Emphasis on Spline Functions", 223-277 (Ed. I.J. Schoenberg) Academic Press.

7. Gordon, W.J. and Riesenfeld, R.F. (1974). Bernstein-Bézier Methods for the Computer-Aided Design of Free Form Curves and Surfaces, *J. of ACM* 21, 293-310.

8. Günttner, R. (1974). Beiträge zur Theorie der Operatoren vom Bernsteinschen Typ, Dissertation, Technischen Universitat Clausthal.

9. Lorentz, G.G. (1966). "Approximation of Functions". Holt, Rinehart and Winston.

10. Schoenberg, I.J. (1959). On Variation Diminishing Approximation Methods, *in* "On Numerical Approximation", 249-274 (Ed. R.E. Langer) University of Wisconsin Press.

11. Schoenberg, I.J. (1967). On Spline Functions, with Supplement by T.N.E. Greville", *in* "Inequalities: Proceedings of a Symposium", 255-291 (Ed. O. Shisha) Academic Press, New York.

12. Stancu, D.D. (1963). A Method for Obtaining Polynomials of Bernstein type of Two Variables, *Amer. Math. Monthly*, 15, 260-264.

13. Stancu, D.D. (1964). The Remainder of Certain Linear Approximation Formulas in Two Variables, *J. SIAM Numer. Anal.* Ser.B, 1, 137-162.

14. Stancu, D.D. (1969). A New Class of Uniform Approximating Polynomial Operators in Two and Several Variables, *in* "Proceedings of the Conference on Constructive Theory of Functions", 443-455, Budapest.

15. Stancu, D.D. (1970). Probabilistic Methods in the Theory of Approximation of Functions of Several Variables by Linear Positive Operators, *in* "Approximation Theory", 329-342 (Ed. A. Talbot) Academic Press.

16. Stancu, D.D. (1972). Approximation of Functions by Means of Some New Classes of Positive Linear Operators, *in* "Numerische Methoden der Approximationstheorie, 187-203 (Eds. I. Collatz and G. Meinardus) ISNM 16, Birkhäuser Verlag.

17. Stancu, D.D. (1970). "Approximation Properties of a Class of Linear Positive Operators", Studia Univ. Babes-Bolyai, Cluj, 15, 33-38.

18. Stancu, D.D. (1971). On the Remainder of Approximation of Functions by Means of a Parameter-dependent Linear Polynomial Operator, *ibid.* 16, 59-66.

19. Stancu, D.D. (1976). Use of Linear Interpolation for Constructing a Class of Bernstein Polynomials, *Studii Cercet. Matem.* 28, 369-379.

20. Strang, G. (1962). Polynomial Approximation of Bernstein Type, *Trans. Amer. Math. Soc.* 105, 525-535.

21. Steffensen, J.F. (1927). "Interpolation", Williams and Wilkins Co.

REMARKS ON THE CONSTRUCTION OF SCHAUDER BASES AND ON THE PÓLYA ALGORITHM

L.A. Karlovitz*

Institute for Physical Sciences and Technology
University of Maryland
College Park, MD 20742

1. INTRODUCTION

It is our purpose to state some results on two classical topics in approximation theory: The construction of Schauder bases for representing functions in infinite dimensional spaces and the construction of best supremum norm approximations of functions by linear families. We note that for Hilbert spaces the analogous problems of constructing orthonormal bases and constructing nearest points rely on the same geometric analysis. However, for non-Hilbert spaces they require quite different considerations. We do not draw any explicit connections between the topics in the sequel.

2. A GEOMETRIC CONSTRUCTION OF SCHAUDER BASES

Much of approximation theory relies on the interplay between the geometry of normed spaces and the analysis of the functions which constitute them. A classical example is the use of extreme points in the derivation of useful characterisations of best approximations for certain classes of functions. In this spirit we shall show that, for a class of

---

*Research supported in part by the National Science Foundation.

spaces, a characteristic curve lying in the surface of the unit ball leads directly to the construction of a Schauder basis. We begin with some examples.

Consider the function $g: [0,2] \to C[0,1]$ defined by

$$g(s)(t) = 1 - 2|s/2 - t| \quad .$$

Its range is a curve lying in the surface of the unit ball of $C[0,1]$, having antipodal end points, i.e. $g(0) = -g(2)$. Moreover, the curve is extremal. Its length, measured in the supremum norm metric, is 2. (The function g is its arc length parametrization.) It follows readily that it is the shortest curve connecting its end points, and, in fact, no two antipodal points in the surface of the unit ball can be connected by a shorter curve. For each point we choose a linear functional $g^*(s)$ which supports the unit ball at $g(s)$, i.e. $\|g^*(s)\| = \langle g^*(s), g(s)\rangle = 1$. Clearly $g^*(s)$ is given by the point evaluation functional at s. Now define the functions: $e_o = g(0)$ and

$$e_k = 2^{m-1}\left[-g\left(\frac{2(k-2^m)}{2^m}\right) + 2g\left(\frac{2(k-2^m)+1}{2^m}\right) - g\left(\frac{2(k-2^m)+2}{2^m}\right)\right] ,$$

$$k = 1,2,..., \tag{2.1}$$

where the integer m satisfies $2^m \le k \le 2^{m+1} - 1$. Also define

$$f^*_o = g^*(0) , \quad f^*_k = g^*\left(\frac{2(k-2^m)+1}{2^m}\right) , \quad k = 1,2,..., \tag{2.2}$$

with m as above.

A graph of the functions readily reveals that $1-g(0), e_o, e_1, ...$ is the *classical polygonal basis of* $C[0,1]$ constructed by Schauder[14] in his original paper on the subject. Equivalently, $e_o, e_1, \ldots$ is the polygonal Schauder basis of the subspace of codimension 1 consisting of functions safisfying $f(0) = -f(1)$. Furthermore, the *coordinate functionals* $\lambda_k(\cdot)$, i.e. $h = \Sigma\lambda_k(h)e_k$, of the latter basis are given by $\lambda_o(h) = \langle h, f_o^* \rangle$ and

$$\lambda_k(h) = \langle h, f_k^* \rangle - \sum_{j=0}^{k-1} \lambda_j(h) \langle e_k, f_k^* \rangle , \quad k = 1,2,\ldots . \tag{2.3}$$

We now apply the same construction to $L_1[0,1]$. Choose $u \in L_1[0,1]$ with $u \neq 0$, a.e., and $\|u\| = 1$. Let

$$\phi(r) = \int_0^r |u(t)| \, dt .$$

Define $g: [0,2] \to L_1[0,1]$ by

$$g(s)(t) = \begin{cases} u(t) , & 0 \le t < \phi^{-1}(s/2) , \\ u(t) , & \phi^{-1}(s/2) \le t < 1 . \end{cases}$$

The range of this function is also a curve lying in the surface of the unit ball having antipodal end points. Moreover, it has length 2 and is thus extremal in the same sense as the curve in the previous example. If we choose $u(t) \equiv 1$, then the construction (2.1) yields the *classical Haar system basis* of $L_1[0,1]$: $e_o = u$ and

$$e_k(t) = \begin{cases} -2^m , & (k-2^m)/2^m \le t < (2(k-2^m)+1)/2^{m+1} , \\ 2^m , & (2(k-2^m)+1)/2^{m+1} \le t < (2(k-2^m)+2)/2^{m+1} , \\ 0 , & \text{elsewhere} , \end{cases}$$

for $k \ge 1$, with m as above. If we let $u(t) = 2t$, then the construction (2.1) yields the nonclassical system: $e_o = u$ and

$$e_k(t) = \begin{cases} -2^{m+1}t\,, & [(k-2^m)/2^m]^{\frac{1}{2}} \le t < [(2(k-2^m)+1)/2^{m+1}]^{\frac{1}{2}}\,, \\ 2^{m+1}t\,, & [(2(k-2^m)+1)/2^{m+1}]^{\frac{1}{2}} \le t < [(2(k-2^m)+2)/2^{m+1}]^{\frac{1}{2}}\,, \\ 0\,, & \text{elsewhere}\,, \end{cases}$$

for $k \ge 1$, with m as above. These sawtooth functions also form a Schauder basis of $L_1[0,1]$. (They are not 'orthogonal' as are the Haar functions.) Indeed, for each choice of u the construction (2.1) yields a Schauder basis of $L_1[0,1]$. This follows from the following theorem which puts the construction into a more general setting.

THEOREM. *Let* X *be a Banach space. Suppose* $g\colon [0,2] \to X$ *satisfies*

$$\|g(s)\| = 1 \text{ for each } s \in [0,2]\,,\quad g(0) = -g(2)\,, \quad \text{and } g \text{ is Lipschitz continuous with constant } 1\,. \tag{2.4}$$

Suppose that g *also satisfies*

$$\Big\| \sum_{\substack{i=1 \\ i\ne j,j+1}}^{n} \lambda_i \frac{g(s_i)-g(s_{i-1})}{s_i-s_{i-1}} + (\lambda_j+\lambda_{j+1}) \frac{g(s_{j+1})-g(s_{j-1})}{s_{j+1}-s_{j-1}} \Big\| \le \Big\| \sum_{i=1}^{n} \lambda_i \frac{g(s_i)-g(s_{i-1})}{s_i-s_{i-1}} \Big\|\,, \tag{2.5}$$

for real λ_i *and* $0 \le s_o < \ldots < s_n \le 2$. *Then* $e_o, e_1, \ldots, e_n, \ldots$ *constructed according to (2.1) yields a Schauder basis for the closed linear hull of* $\{g(s)\colon s \in [0,2]\}$. *Moreover, the component functionals are given by (2.3).*

The hypotheses are readily verified for each of the examples above. *Thus we have a geometrically motivated common construction for Schauder bases for a large class of spaces, including the classical polygonal basis of* $C[0,1]$ *and the Haar system as well as a multitude of nonclassical bases of* $L_1[0,1]$. Apropos the general class we note that any Banach space X for which there exists a function $g: [0,2] \to X$ satisfying (2.4) is termed a *flat Banach space*. This class has been introduced by Harrell-Karlovitz[2,3,4] and forms a large proper subset of all nonreflexive spaces. The term *flat* follows from their distinguished geometric properties. A variety of examples, including (isomorphically) the new space of Lindenstrauss-Stegall[7], are treated in [2,3,4], Nyikos-Schäffer[8], and Schäffer[12,13]. As indicated herein, the structure of flat spaces can be usefully exploited for the study of various problems of analysis and geometry.

The minimum length property of the curves discussed here can be expressed in terms of an interesting geometric parameter of normed spaces, the *girth of the unit ball*, introduced by Schäffer[11]. Briefly, a curve satisfying (2.4), together with its reflection through the origin, forms a centrally symmetric curve lying in the surface of the unit ball, which because of its minimal length achieves the girth of the unit ball (= 4) of the space. Study of this parameter has led to a variety of interesting results and questions in finite as well as infinite dimensional spaces. We refer to Schäffer[13] for a new discussion of girths, flat spaces, and related topics.

The construction does not yield a Schauder basis for all flat Banach spaces, for it is known to fail in some situations where (2.5) is not satisfied. On the other hand, the construction is known to yield a Schauder basis in some situations wherein (2.4) is only partially satisfied. Thus, in particular, it can be applied to certain non-flat spaces. The proof of the theorem as well as other versions will be given elsewhere.

3. THE PÓLYA ALGORITHM

The classical Pólya 'algorithm' is concerned with approximating the best supremum norm approximation of a continuous function by best L^p norm approximations. Let Ω be a compact Lebesgue measurable subset of R^n and $C(\Omega)$ the space of all real-valued (or complex valued) continuous functions on Ω. Let U be a finite dimensional linear subspace of $C(\Omega)$, and fix $f \in C(\Omega)$. Let $G(p)$ denote the best approximation of f by U in the L^p norm. Then $\{G(p): 0 < p < \infty\}$ is bounded in the supremum norm and each convergent subsequence, with $p \to \infty$, has for its limit a best supremum norm approximation of f by U.

Since, in general, each $G(p)$ can only be constructed as the limit of an infinite process, the 'algorithm' is doubly infinite. In Karlovitz[5] we have given a singly infinite version of the algorithm which proceeds with approximations of the $G(p)$. The latter has two basic limitations. The first is inherited from the Pólya algorithm, wherein it may happen (Descloux[1]) that different convergent subsequences have different limits. Thus an algorithm (singly or doubly infinite) which insures convergence for the entire sequence for each continuous function has yet to be devised. Interesting related questions are to characterize those functions for which the Pólya algorithm converges, and to devise averaging methods which will insure convergence. The second shortcoming of the algorithm in [5] is the hypothesis that $f(x)-g(x) \neq 0$, a.e. for each $g \in U$. Thus the extension of the algorithm to atomic measures, e.g. to finite Ω, is not immediate.

We propose to state here a version of the algorithm which is designed for finite Ω and which has the additional feature that the entire sequence converges for each function defined on Ω.

Let $\Omega = \{x_1,\ldots,x_m\}$. For each $g \in U$ and $p \geq 2$ we define the weight function $w = w(g,p)$

$$w(x_i) = \begin{cases} \varepsilon(p)^p , & \text{if } f(x_i) = g(x_i) , \\ |f(x_i)-g(x_i)|^{p-2} , & \text{if } f(x_i) \neq g(x_i) , \end{cases}$$

where $\varepsilon(p)$ is chosen so that $\varepsilon(p) > 0$ and $\varepsilon(p) \to 0$ as $p \to \infty$.

We construct a sequence (g_k,h_k,p_k) with $g_k,h_k \in U$ and $2 \leq p_k < \infty$. To start, choose $g_1 \in U$ and $p_1 \geq 2$. Then h_1 is the unique solution of the weighted L_2-approximation

$$\min \|f-g\|^2_{w(g_1,p_1)} = \min \sum_{i=1}^{m} w(x_i)\ |f(x_i)-g(x_i)|^2 , \quad g \in U .$$

If (g_k,h_k,p_k) has been constructed, then $g_{k+1} = g_k+\lambda_o(h_k-g_k)$, where λ_o minimizes the convex function

$$\min \|f-g_k-(h_k-g_k)\|_{L^{p_k}} , \quad 0 \leq \lambda \leq 1 .$$

Finally we consider the quantity

$$\|f-g_k\|^2_{w(g_k,p_k)} - \|f-h_k\|^2_{w(g_k,p_k)} - \varepsilon(p)^p .$$

If this is negative, we choose $p_{k+1} > p_k$. If it is positive, we let $p_{k+1} = p_k$. Then we have:

THEOREM. *The sequence* $g_1,g_2,\ldots,g_k,\ldots$ *constructed in this manner converges to* $g_o \in U$ *which is a best supremum norm approximation of* f *by* U. *The function* g_o *can be defined independently; namely, it is the so-called 'strict' best supremum norm approximation (in the sense of Rice) of* f *by* U.

A discussion of the notion of 'strict' best approximation appears in Descloux[1]. As stated, this represents a theoretical algorithm. It is, however, open to a number of straightforward modifications and convergence acceleration schemes. To date, only a limited amount of numerical work has been done.

The algorithm has a point of contact with the 'Lawson algorithm', which is discussed in Rice-Usow[10], in its use of weighted L_2-approximations towards the construction of best supremum norm approximations. However, the motivations and details are quite different. The Lawson algorithm is based on a result of Motzkin and Walsh dealing with the weighted approximation of functions of one variable by Chebyshev families and does not make use of the Pólya algorithm. The algorithm appears to be limited to approximation by Chebyshev families of functions defined on finite sets. Rice and Usow have extended the algorithm to the use of weighted L^p, p fixed, approximations for $p \neq 2$.

The proof of the theorem as well as further discussion, including extensions to convex families of functions, will be given elsewhere.

REFERENCES

1. Descloux, J. (1963). Approximation in L^p and Chebyshev approximations, *J. Soc. Indust. Appl. Math.* 11, 1017-1026.

2. Harrell, R.E. and Karlovitz, L.A. (1970). Girths and flat Banach spaces, *Bull. Amer. Math. Soc.* 76, 1288-1291.

3. Harrell, R.E. and Karlovitz, L.A. (1974). The geometry of flat Banach spaces, *Trans. Amer. Math. Soc.* 192, 209-218.

4. Harrell, R.E. and Karlovitz, L.A. (1975). On tree structures in Banach spaces, *Pacific J. Math.* 59, 85-91.

5. Karlovitz, L.A. (1970). Construction of nearest points in the L^p, p even, and L^∞ norms, *J. Approx. Theory* 3, 123-127.

6. Karlovitz, L.A. (1976). Construction of a Schauder basis for a class of nonreflexive Banach spaces, Univ. of Maryland Tech. Note.

7. Lindenstrauss, J. and Stegall, C. Examples of spaces which do not contain ℓ_1 and whose duals are separable, *Studia Math.*

8. Nyikos, P. and Schäffer, J.J. (1972). Flat spaces of continuous functions, *Studia Math.* 42, 221-229.

9. Pólya, G. (1913). Sur un algorithme toujours convergent pour obtenir les polynomes de meilleure approximation de Tchebycheff pour une fonction continue quelconque, *Comp. Rend.* 157, 480-483.

10. Rice, J.R. and Usow, K.H. (1969). Lawson's algorithm and extensions, *Math. Comp.* 22, 118-127.

11. Schäffer, J.J. (1967). Inner diameter, perimeter, and girth of spheres, *Math. Ann.* 173, 59-79.

12. Schäffer, J.J. (1971). On the geometry of spheres in L-spaces, *Israel J. Math.* 10, 114-120.

13. Schäffer, J.J. (1976). "Geometry of Spheres in Normed Spaces", Marcel Dekker, New York.

14. Schauder, J. (1927). Zur Theorie stetiger Abbildungen Funktionalräumen, *Math. Z.* 26, 47-65.

ON A CLASS OF METHODS FOR NONLINEAR APPROXIMATION PROBLEMS

G.A. Watson

University of Dundee
Dundee, Scotland

1. INTRODUCTION

For the fitting of a nonlinear model to discrete data, the optimal parameters are typically chosen to minimise the least squares norm, and a standard method in this case is the Gauss-Newton algorithm. In recent years, generalisations of this approach have been given which apply when the criterion used is more appropriately some other norm: in particular, applications to the L_∞ norm are given in [4] and [7], and to the L_1 norm in [8]. The further generalisation of these results to polyhedral norms is considered in [1], and in [9] the analysis is extended to include all norms in R^m.

It is the purpose of this paper to consider the application of the Gauss-Newton method in a completely general (real) setting. In particular, the removal of dimensionality restrictions means that we are able to include the treatment of problems involving continuous real-valued functions of many variables. The general problem considered here may be stated:

$$\text{find } \underset{\sim}{a} \in R^n \text{ to minimise } \|f\| \quad , \tag{1.1}$$

where $f(\underset{\sim}{a}) : R^n \to M$, with M a real, linear space equipped with norm $||\cdot||$. We will show that the limit points of the modified Gauss-Newton iteration are, under mild assumptions, stationary points of $||f||$: the appropriate characterisation of the stationary points is given in Section 2, and in Section 3 we give the details of the method, and a basic convergence result. Finally in Section 4, we draw attention to certain fundamental differences in the performance of the algorithm which depend on the nature of the norm used.

The analysis presented here is essentially a local one, and we will restrict attention to problems for which (i) there exists a bounded region $B \subset R^n$ containing a stationary point, (ii) f is sufficiently smooth in B that we can write

$$f(\underset{\sim}{a}+\underset{\sim}{d}) = f(\underset{\sim}{a}) + \ell(\underset{\sim}{d}) + ||\underset{\sim}{d}||_A^2 w(\underset{\sim}{a},\underset{\sim}{d}) \quad , \qquad (1.2)$$

with $||w|| \leq W$ in B . Here $\ell(\underset{\sim}{d})$ denotes the linear combination $\sum_{j=1}^{n} d_j g_j(\underset{\sim}{a})$, where g_j is the partial derivative of f with respect to a_j , $j = 1,2,\ldots,n$, and $||\cdot||_A$ is any norm on R^n . For example, (ii) is satisfied if f is a twice continuously differentiable mapping of B into M . It will be assumed that the algorithm produces iterates which are confined to B .

2. CHARACTERISATION OF STATIONARY POINTS

Let $\langle\cdot,\cdot\rangle$ be an inner product between the elements of M and those of the dual space M^* . Then we can write

$$||u|| = \sup_{||v||^* \leq 1} \langle u,v\rangle \quad , \qquad (2.1)$$

where $||\cdot||^*$ is the dual norm on M^* . It is convenient to denote by $V(u)$ the (convex) set of elements of M^* at which the supremum in (2.1) is attained, so that

$$V(u) = \{v : \|u\| = \langle u,v\rangle \; , \; \|v\|^* \leq 1\} \quad . \tag{2.2}$$

Then we have the following result (which may also be obtained from the analysis given in [2]).

THEOREM 2.1 If $\underset{\sim}{a}$ minimises $\|f\|$, $\exists\; v \in V(f(\underset{\sim}{a}))$ such that $\langle g_j, v\rangle = 0$, $j = 1,2,\ldots,n$.

Proof Let $\underset{\sim}{a}$ minimise $\|f\|$, but assume that the conditions are not satisfied. Then, by the theorem on linear inequalities ([3], p.19) $\exists\; \underset{\sim}{d} \in R^n$, $\delta > 0$ such that

$$\langle \ell(\underset{\sim}{d}), v\rangle < -\delta, \forall\; v \in V(f(\underset{\sim}{a})) \quad .$$

Now, for any $v(\gamma) \in V(f(\underset{\sim}{a}+\gamma\underset{\sim}{d}))$

$$\|f(\underset{\sim}{a}+\gamma\underset{\sim}{d})\| = \langle f(\underset{\sim}{a}+\gamma\underset{\sim}{d}), v(\gamma)\rangle$$

$$< \langle f(\underset{\sim}{a}), v(\gamma)\rangle - \gamma\delta + \gamma\langle \ell(\underset{\sim}{d}), v(\gamma) - v\rangle + O(\gamma^2) \; . \tag{2.3}$$

By the weak* compactness of the unit ball in M^* (Alaoglu-Bourbaki theorem, e.g. Holmes[5]), $\exists$ a positive sequence $\{\gamma_j\} \to 0$ and $m \in M^*$ such that $\langle u, v(\gamma_j) - m\rangle \to 0$, $j \to \infty$, for all $u \in M$. Further

$$0 \leq \|f(\underset{\sim}{a})\| - \langle f(\underset{\sim}{a}), v(\gamma)\rangle \leq O(\gamma)$$

and so $m \in V(f(\underset{\sim}{a}))$. Letting $\gamma \to 0$ in inequality (2.3) along the sequence $\{\gamma_j\}$, we obtain a contradiction that $\underset{\sim}{a}$ gives a minimum, and the result follows.

Definition 2.1 We say that $\underset{\sim}{a}$ is a *stationary point* of $\|f\|$

if $\exists\ v \in V(f(\underset{\sim}{a}))$ such that $\langle g_j, v\rangle = 0$, $j = 1,2,\ldots,n$.

Remark If $f(\underset{\sim}{a})$ is linear in $\underset{\sim}{a}$, the conditions of Theorem 2.1 are easily seen to be also sufficient for $\underset{\sim}{a}$ to minimise $\|f\|$. Particular choices of normed space (and thus set V) give rise to familiar characterisation results.

3. THE ALGORITHM

Methods of Gauss-Newton type are iterative procedures for the nonlinear problem which are based on obtaining directions along which $\|f\|$ can be reduced by minimising the appropriate norm of a linear approximation to f at the current point. The methods differ in the way in which the step lengths in the descent directions are chosen, and the particular procedure given here is only one of a number of possible strategies. Let $\underset{\sim}{a}_i \in B$ be the current approximation, let $f_i = f(\underset{\sim}{a}_i)$, and $\ell_i(\underset{\sim}{d}) = \sum_{j=1}^{n} d_j g_j(\underset{\sim}{a}_i)$. We consider the following algorithm.

(1) Determine $\underset{\sim}{d}_i \in R^n$ to minimise $\|r\|$, where

$$r = f_i + \ell_i(\underset{\sim}{d})$$

and let

$$r_i = f_i + \ell_i(\underset{\sim}{d}_i) \quad . \tag{3.1}$$

(2) If $\|f_i\| = \|r_i\|$ then stop; otherwise choose γ as the largest element in the set $\{1,\theta,\theta^2,\ldots\}$, $0 < \theta < 1$, so that

$$\psi(\underset{\sim}{a}_i,\gamma) = \frac{\|f_i\| - \|f(\underset{\sim}{a}_i+\gamma\underset{\sim}{d}_i)\|}{\gamma(\|f_i\| - \|r_i\|)} \geq \sigma > 0 \quad , \tag{3.2}$$

where σ is a constant, independent of i, satisfying $\sigma < 1$.

(3) Set $\underset{\sim}{a}_{i+1} = \underset{\sim}{a}_i + \gamma\underset{\sim}{d}_i$, increase i by one, and return to step 1.

That this algorithm can be implemented, and has a basic convergence property, is the substance of the following theorem.

THEOREM 3.1 Provided that $\|\underset{\sim}{d}_i\|_A$ is bounded, the algorithm produces a sequence $\{\underset{\sim}{a}_i\}$ whose limit points are stationary points of $\|f\|$.

Proof Clearly $\|f_i\| \geq \|r_i\|$. Precisely as in [9], it may be shown that if $\|f_i\| > \|r_i\|$, $\exists\ \gamma > 0$ satisfying (3.2), and the sequence of points produced is such that $\|f_i\| - \|r_i\| \to 0$, $i \to \infty$. Thus the scalar equation $\|f\| = \|r\|$ will be satisfied at the limit points of the sequence $\{\underset{\sim}{a}_i\}$. Now the condition that $\|r\|$ is minimised is that $\exists\ s \in V(r)$ with $\langle g_j, s\rangle = 0$, $j = 1,2,\ldots,n$. Thus, at any limit point $\underset{\sim}{a}$

$$\|f\| = \|r\| = \langle r,s\rangle = \langle f,s\rangle$$

and so $s \in V(f(\underset{\sim}{a}))$, showing that a is a stationary point of $\|f\|$.

4. CONVERGENCE OF THE FULL STEP METHOD

The usefulness of the algorithm will depend on a reasonable rate of convergence being achieved, and that is unlikely unless the step lengths γ are bounded away from zero. In particular, conditions on the problem which guarantee convergence of the full step method ($\gamma = 1$) have been obtained for a number of different norms in R^m . It was first pointed out by Osborne[6] that the performance of the Gauss-Newton method can differ substantially according to the norm used. He contrasted the L_∞ and L_2 cases, showing in particular that the L_∞ norm iteration could have a faster rate of convergence than was possible in the L_2 case. The behaviour of the method in these two special cases in fact typifies the behaviour in two wider classes of norm.

In order for the full step method to converge, we must have from (3.2)

$$\psi(\underset{\sim}{a}_i, 1) \geq \sigma \quad , \quad \forall\ i \quad , \tag{4.1}$$

and, using equation (1.2), it is readily seen that this will be satisfied provided that

$$\|\underset{\sim}{d}_i\|_A^2 \leq \frac{1-\sigma}{W} (\|f_i\| - \|r_i\|) \quad , \quad \forall\ i \quad . \tag{4.2}$$

For polyhedral norms in R^m, conditions are given in [1] for the inequality

$$\|d_i\|_A^2 \leq K(\|f_i\| - \|r_i\|)^2$$

to be satisfied for all $\underset{\sim}{a}_i$ in a neighbourhood of the appropriate limit point, ensuring that ultimately the full step method is convergent. It is also shown that the rate of convergence is second order. When $M = C[X]$, X a compact set, Cromme[4] has obtained a similar rate of convergence result for the L_∞ norm, provided that a closely related condition of 'local strong uniqueness' is satisfied.

For problems involving norms from the class of smooth, strictly convex, monotonic norms, the behaviour of the Gauss-Newton method is quite different. This difference can be interpreted as a consequence of the weaker inequality which is obtained for $\|\underset{\sim}{d}_i\|_A$, and is illustrated by the following result (in which the subscripts i have been omitted).

THEOREM 4.1 Let $M = C[X]$, X an N-dimensional continuum, $\|\cdot\| = \|\cdot\|_p$, $2 \le p < \infty$. Let $E = \{x \in X : |r| \ge \|r\|_\infty/2\}$ and assume that

(i) $$\min_{\|\tilde{t}\|_A = 1} \int_E (\ell(\tilde{t}))^2 dx \ge \delta > 0$$

(ii) $$\|r\| \le K\|r\|_\infty \quad .$$

Then

$$\|d\|_A^2 \le \frac{2(2K)^{p-2}}{\delta} \|f\| \, (\|f\| - \|r\|) \quad . \tag{4.4}$$

Proof The condition that $\|r\|$ is a minimum is, by Theorem 2.1,

$$\langle g_j, s\rangle = 0 \quad , \quad j = 1,2,\ldots,n \quad , \tag{4.5}$$

for some $s \in V(r)$. Here the inner product is given by $\langle r,s\rangle = \int_X rs\,dx$, with $s = r|r|^{p-2}\|r\|^{1-p}$. Using (4.5) we have the identity

$$\langle \ell(d)|r|^{p-2}, \ell(\tilde{d})\rangle = \langle f|r|^{p-2}, f\rangle - \langle r|r|^{p-2}, r\rangle$$

$$\le \|r\|^{p-2}(\|f\|^2 - \|r\|^2)$$

using Holder's inequality

$$\le 2\|r\|^{p-2} \|f\| \, (\|f\| - \|r\|) \quad .$$

Letting $t \in R^n$ with $\|\tilde{t}\|_A = 1$, the result follows.

The essential requirement for (4.4) is that the set $\{g_j\}$ be linearly independent on X. This is weaker than is required in the corresponding discrete case, when a similar inequality follows if the $m \times n$ matrix A of partial derivatives satisfies the Haar condition [9]. An exception is when $p = 2$, when the requirement is just that A have full rank.

Assuming that constants K and δ exist which are independent of the iteration, the full step method is therefore convergent in this case only if $\|f\|$ becomes sufficiently small. Another difference occurs in the convergence rate. Let $c_j(\underset{\sim}{a}) = g_j |r|^{p-2}$, $j = 1,2,\ldots,n$, where r is the element of minimum norm.

THEOREM 4.2 Let $\{c_j(\underset{\sim}{a})\}$ be continuously differentiable functions of $\underset{\sim}{a}$ in B. Then the rate of convergence of the full step method is first order.

Remark The conditions are satisfied if the functions g_j are smooth, and $p = 2$ or $3 < p < \infty$.

Proof Let $\underset{\sim}{a}^*$ be a stationary point. Then

$$f(\underset{\sim}{a}^*) = f_i + \ell_i(\underset{\sim}{a}^* - \underset{\sim}{a}_i) + \|\underset{\sim}{a}^* - \underset{\sim}{a}_i\|^2 w_i$$

$$= r_i + \ell_i(\underset{\sim}{a}^* - \underset{\sim}{a}_{i+1}) + \|\underset{\sim}{a}^* - \underset{\sim}{a}_i\|^2 w_i \quad .$$

Thus

$$\langle \ell_i(\underset{\sim}{a}^* - \underset{\sim}{a}_{i+1}), c_j(\underset{\sim}{a}_i) \rangle = \langle f(\underset{\sim}{a}^*), c_j(\underset{\sim}{a}_i) - c_j(\underset{\sim}{a}^*) \rangle$$

$$- \|\underset{\sim}{a}^* - \underset{\sim}{a}_i\|^2 \langle w_i, c_j(\underset{\sim}{a}_i) \rangle \quad , \quad j = 1,2,\ldots,n \ ,$$

where we have used (4.5) at $\underset{\sim}{a}_i$ and $\underset{\sim}{a}^*$. If $\|r_i\| > 0$, the $n \times n$ matrix with (k,j) element $\langle g_k, c_j \rangle$ is positive definite if the set $\{g_k\}$ is linearly independent, and we obtain the required result.

COROLLARY If $f(\underset{\sim}{a}^*) = 0$, then the rate of convergence is second order.

ACKNOWLEDGEMENT

I am grateful to B. Brosowski, L.A. Karlovitz and J. Meinguet for valuable comments.

REFERENCES

1. Anderson, D.H. and Osborne, M.R. Discrete, Non-Linear Approximation Problems in Polyhedral Norms, *Num. Math.* (in preparation).

2. Brosowski, B. and Wegmann, R. (1970). Charakterisierung bester Approximationen in normierten Vektorräumen, *J. Approx. Theory* 3, 369-397.

3. Cheney, E.W. (1966). Introduction to Approximation Theory, New York, McGraw-Hill.

4. Cromme, L. (1976). Eine Klasse von Verfahren zur Ermittlung bester nicht-linearer Tschebyscheff-Approximationen, *Num. Math.* 25, 447-459.

5. Holmes, R.B. (1975). Geometric Functional Analysis and its Applications, Springer-Verlag, New York.

6. Osborne, M.R. (1972). An Algorithm for Discrete, Non-Linear Best Approximation Problems, *in* "Numerische Methoden der Approximationstheorie, Band 1 (Eds. L. Collatz and G. Meinardus) Birkhauser Verlag.

7. Osborne, M.R. and Watson, G.A. (1969). An Algorithm for Minimax Approximation in the Non-Linear Case, *Computer J.* 12, 64-69.

8. Osborne, M.R. and Watson, G.A. (1971). On an Algorithm for Nonlinear L_1 Approximation, *Computer J.* 14, 184-188.

9. Osborne, M.R. and Watson, G.A. (1978)."Nonlinear Approximation Problems in Vector Norms", Numerical Analysis Dundee 1977 Proceedings (Ed. G.A. Watson).

SOME PRACTICAL APPLICATIONS IN DISCRETE MULTIVARIATE APPROXIMATION

J.G. Hayes

Division of Numerical Analysis and Computing
National Physical Laboratory
Teddington, Middlesex

ABSTRACT

This paper describes a number of practical applications of multivariate approximation, in two up to ten or so independent variables, and discusses the considerations which arise in such problems over and above the question of providing sound numerical algorithms. Chief among these considerations is the need to construct a suitable functional form with which to carry out the approximation.

1. INTRODUCTION

Work in multivariate approximation presented at this conference covers a wide spectrum. We see reference to work on entropy and widths, optimal recovery, remainder functionals, existence, uniqueness and characterization, and construction of algorithms. This work has varying degrees of practical motivation, but the last category in particular aims at providing the means to solve practical problems. Instead of discussing algorithms in this paper, however, my intention is to try to complete this end of the spectrum by describing a number of specific applications. This is partly in the hope that they will provide some interest in themselves, but also to make two particular points. These are (a) that having a good algorithm is not the only consideration when seeking an approximation in practice, and (b) that the other considerations which arise dominate the solution process more and more as the number of dimensions of the problem increases. The applications I shall describe involve from two up to ten or so independent variables. They are a selection from those I have encountered at various times over quite a number of years, a selection which I shall restrict as far as possible to published material, so that further details are readily available, if desired.

The most important of these additional considerations is the need to choose an appropriate form of function with which to carry out the approximation. In univariate approximation, in the usual absence of any counter-indication from the theoretical context of the problem, part of that choice will be automatic: we shall use either a general polynomial or a cubic spline. The rest of the choice will consist of (possibly) an initial transformation of variables and finally the degree of the polynomial or the number and positions of the knots. The same applies to a large extent in bivariate approximation, though the choice is often very much more difficult. In higher dimensions the choice is, in one respect, more restricted, since the spline when extended to more variables (see Hayes and Halliday[9] for example) soon becomes too cumbersome: in six variables for instance a cubic spline with only one interior knot in each variable has over 15,000 coefficients. Thus we are usually confined to polynomials, indeed polynomials of quite low degree, or at any rate with a comparatively small number of terms (which clearly will limit the problems we can solve). This is at least partly due to the fact that many, probably most, practical applications are concerned with discrete data, and the number of data points often puts a severe limit on the number of coefficients that can be used. Even so, we shall see that the choice we have to make is often a difficult one.

Perhaps it should be remarked, finally, that many problems in many dimensions arise in which the fitting function is simply a linear combination of the variables. If the set of variables is given, there is no choice to make. If the problem is to select a good set of variables from a larger set, a method such as that of Efroymson[3] is appropriate. In the present paper, however, we are concerned with situations which are distinctly nonlinear.

2. TRANSISTOR VOLTAGE (2 independent variables)

The first application is one in which it certainly is important to have the right kind of fitting algorithm. It is a fairly straightforward polynomial surface problem which arose recently in research into active electronic filters at an industrial laboratory. In manufacture, transistors are formed in a regular matrix on a silicon slice, and part of the investigation was to model the way a particular junction voltage varied with the position of the transistor on the slice. So voltage values are given at various positions in the (x,y) plane, and to these we decide to fit a polynomial surface by least-squares.

The main feature to be desired in the fitting algorithm, in addition to numerical stability, is that it should express the fitted polynomial $f(x,y)$ in terms of Chebyshev polynomials. There are two forms in which it might reasonably do this, namely

$$f(x,y) = \sum_{i=0}^{k} \sum_{j=0}^{l} a_{ij} T_i(x) T_j(y) \quad , \qquad (1.2)$$

and

$$f(x,y) = \sum_{i=0}^{i+j=k} \sum_{j=0} a_{ij} T_i(x) T_j(y) \quad , \qquad (2.2)$$

where $T_i(x)$ and $T_j(y)$ are Chebyshev polynomials of the first kind of degree i and j respectively, and it is assumed that x and y have both been normalized so that their data values span the range -1 to $+1$. The form (2.1), allowing different degrees in the two variables, is the more generally useful, since with disparate entities, pressure and temperature say, there is no reason to expect the same degree in each. In the present application, however, where both variables are distances over a physical plane and the variation of voltage is expected to have the same character in both directions, it is reasonable to use form (2.2).

The numerical values of the coefficients of the least-squares polynomial of degree 7 are given in Table I, arranged

Table I Coefficients a_{ij} in form (2.2) with $k = 7$

j/i	0	1	2	3	4	5	6	7
0	+477	−306	−157	+ 80	+ 57	+ 47	+ 11	+ 1
1	− 30	−308	− 99	−125	− 44	− 7	− 10	
2	−153	+219	+173	+180	+ 51	+ 50		
3	− 21	−143	− 15	− 22	+ 2			
4	+ 27	+ 79	+ 45	+ 36				
5	− 7	− 13	+ 23					
6	− 1	− 17						
7	+ 2							

in the natural triangular array. The algorithm used comprised a direct setting-up of the overdetermined linear equations following from form (2.2) and their least-squares solution via orthogonal transformations in the tradition of Golub[4]. The combination of a Chebyshev polynomial basis and orthogonal transformation provides a very stable algorithm. Degree 7 was selected, in the expectation that it would be higher than ultimately required, so as to provide the information to help settle the main question, the degree to choose for the final fit, namely the lowest degree which provides an adequate approximation to the data.

For this purpose, the advantage of the Chebyshev form of the fitted polynomial is that its terms can be expected to decrease more rapidly with increasing degree than any other form, until they settle down to a roughly constant size, about the size of the data errors or rather less. Of course, since the values of Chebyshev polynomials span the range -1 to +1 (for values of their argument between -1 and +1), we need only examine the values of the coefficients to see the maximum effect of each term in the series. This feature is shared by the power-series form of polynomial (in the normalized variables), but the coefficients in this case will decrease much more slowly, and indeed may not decrease at all. In particular, they will not settle down as the Chebyshev coefficients do and so will provide no guidance as to the degree to choose.

Thus we examine the Chebyshev coefficients with a view to detecting the settling down to a constant size and then select the lowest degree which avoids discarding any coefficients significantly larger than this size. The situation closely parallels that in the curve-fitting case, for which an example with a clear-cut decision is given in Hayes[7]. In the curve-fitting case, as can also be seen in the quoted example, equivalent information is provided by the list of root-mean-square residuals s_i, for increasing degree i of the fitted polynomial. Here s_i is defined by

$$(n-p_i)s_i^2 = \sum_{r=1}^{m} \left[f^{(i)}(x_r) - y_r\right]^2 , \qquad (2.3)$$

where $f^{(i)}(x)$ is the fitted polynomial of degree i, p_i is the number of coefficients in the polynomial, and (x_r, y_r), $r = 1,2,\ldots,m$, are the data points. We can compute a corresponding set of rms residuals in the surface case, by building up the polynomial one term at a time and calculating the rms residual at each stage, but the information thus provided is less useful in the surface case than that from the Chebyshev coefficients. For instance, we may start with $k = 0$ in (2.2) and add the linear terms one at a time, then the second degree terms one at a time and so on. By this means we shall obtain an rms residual (as we obtained a Chebyshev coefficient) corresponding to each term in the polynomial, but its value will be dependent on which terms of the same degree have been added previously. Alternatively, we may examine only the rms values for $k = 0,1,2,\ldots$ in (2.2), but then there is less information and we should need to go to a higher degree to detect the settling down. Thus the Chebyshev coefficients are particularly valuable in the surface case.

Our results in Table I do not provide such a clear-cut decision as those of the curve-fitting example, but the coefficients appear to have settled down fairly well to values in a range of about ± 50 (which in fact coincides with the rms value for $k = 7$). On this basis we would choose degree 5 for the polynomial finally to be fitted: it would, however, be necessary to examine the shape of the fitted polynomial in some detail before accepting it.

3. SHIP FAIRING (2 independent variables)

The quality of the fitting algorithm used can also be important in much more complicated applications, and we shall look briefly at the ship-fairing problem. Figure 1 gives a typical body-plan, which shows the transverse vertical section of the hull at various distances along it. The sections being

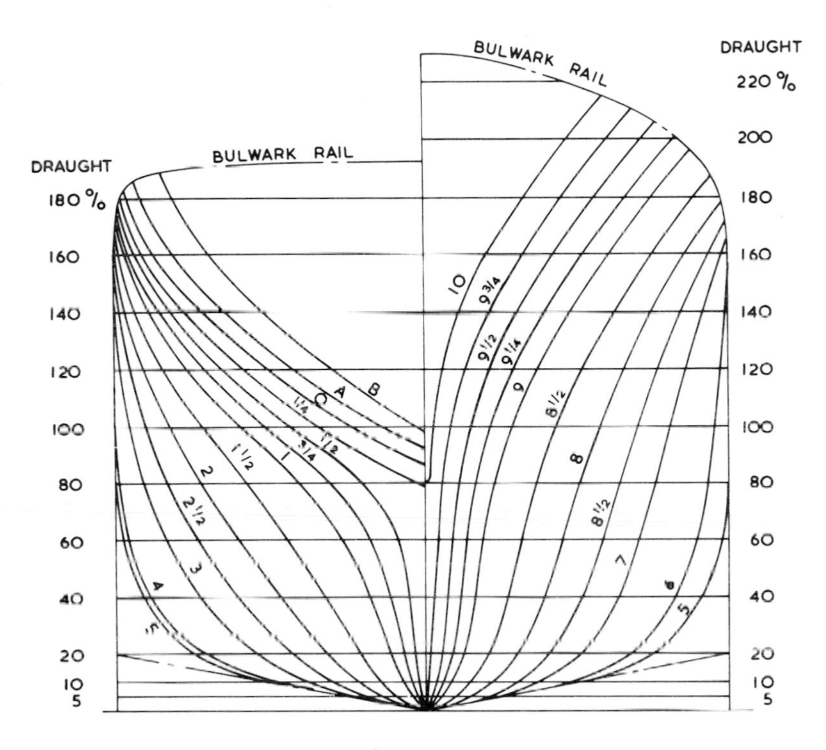

Figure 1

symmetric, only half of each one is shown, those from the forebody of the ship on the right and those from the afterbody on the left. The data, taken from a drawing typically to a scale of 1:50 , consist of the offset, or half-breadth, at each station and each height shown on the figure, together with boundary information. To ensure single-valued functions, we seek a mathematical representation of the hull which expresses the offset as a function of the other two parameters.

We can see at once that the shapes of the curves are not amenable to purely polynomial representation. Many of the sections, particularly in the afterbody, consist of a pair of straight lines joined by a curved portion, the transition from one line to the other being very rapid near the centre section, section 5. The curved portion of this latter section is in fact a circular arc. Section $\frac{1}{2}$ is also non-polynomial in character, often more so than here, with a longer nearly straight portion at its lower end. Moreover, if we were to examine the horizontal sections (waterlines) of the hull, we would see that the higher waterlines have straight portions at the centre and infinite derivatives at the after end, while the lower waterlines are curved throughout, with small derivatives at the after end; and the change from one type of waterline to the other occurs quite rapidly. A further difficulty arises from the way in which the slope of the sections at their lower end varies along the length of the vessel: it is large (in the coordinate system defined) and constant along most of the length, but rapidly gets small at each end of the hull.

It thus becomes necessary to divide the hull surface into a number of separate pieces and, for each one, to provide the capability of reflecting its particular features. At the same time, the separate pieces must join smoothly together. We see, then, that a very versatile fitting algorithm is certainly required, but we see also that the identification of functional forms which have the capability of providing a good fit has become a major part of the investigation. Hayes[5] describes a fitting system for this problem, based on polynomials but with substantial enhancements. We may comment finally that many of the difficulties of the problem can, in principle, be overcome by the use of parametric splines, but in practice the approach described has so far proved much the more successful.

4. LIFTING GEAR (2 and 3 independent variables)

Our next application is one in which, though we are still concerned with only two independent variables, the identification of a suitable functional form constituted essentially the whole investigation. It is, in fact, a mathematical function rather than discrete data that we wish to approximate, and we are particularly interested in maximum error, so we shall use the Chebyshev rather than the l_2 norm - though, as is often the case, the difference is of little practical significance. The application, discussed in Anthony, Gorley and Hayes[1], is concerned with the derivation of a design formula for a basic component of lifting-gear, the chain link. The notation of the link geometry is shown in Figure 2. The

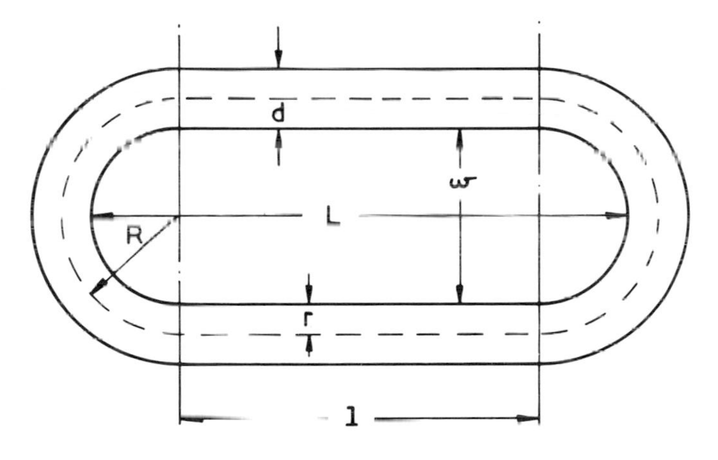

Figure 2

parameters d, w and L are the ones used in practical work, but the simply related-set r, R and l is the more natural for mathematical stress analysis. In addition we have two other parameters, F, the design load, and f, the design stress, which for our purposes has to be equated to the maximum stress in the link.

At the start of the investigation, one design formula had long been available, namely

$$F = \frac{0.224\, fd^3}{w + 0.4d}\left[1.75 + \frac{w+d}{L+d}\right] , \tag{4.1}$$

with an extra multiplicative factor $0.22(2 + L/d)$ whenever $L < 2.55d$. This formula is quoted in British Standards as being accurate to within $\pm 4\%$. In the design use of this formula, however, it is required to compute d given the values of the other parameters, and so a requirement arose for a formula in a more convenient form, namely expressing d directly in terms of the other parameters. To provide a better basis for this purpose than the approximate formula (4.1), one of my co-authors, T.A.E. Gorley, derived the following formula by linear elastic stress analysis:

$$f = \frac{F}{2\pi r^2}\left[\frac{1+r^2/2ZR}{1+\pi r^2(Z+1)/4ZR}\right]\left[1 + \frac{1}{Z}\,\frac{r}{R+r}\right]$$

or (4.2)

$$\frac{F}{2\pi r^2}\left[\frac{1+r^2/2ZR}{1+\pi r^2(Z+1)/4ZR}\right]\left[1 - \frac{1}{Z}\,\frac{r}{R-r}\right] + \frac{F}{2\pi r Z(R-r)} ,$$

whichever is the larger. Here Z is given by

$$Z = \left(\frac{r}{R}\right)^2\left\{\left[1 - \left(\frac{r}{R}\right)^2\right]^{\frac{1}{2}} + 1\right\}^2 . \tag{4.3}$$

That there are two forms in (4.2) reflects the fact that the maximum stress can occur at either of two different positions on the link depending on its geometry. The problem, then, is to determine a formula expressing d in terms of F , f , w

and L which approximates (4.2) to an accuracy better than 4% , hopefully about 2% , over the ranges

$$0.1 \le \frac{r}{R} \le 0.6 \quad \text{and} \quad 0 \le \frac{l}{R} \le 4 \quad . \tag{4.4}$$

The two sets of geometric parameters are related by

$$d = 2r \quad , \quad w = 2R - 2r \quad , \quad L = l + 2R - 2r \quad . \tag{4.5}$$

We may observe, first of all, that, though there are five parameters, it is essentially only a 3-variable problem. For example, (4.2) can be expressed entirely in terms of the three non-dimensional parameters

$$\frac{F}{fL^2} \quad , \quad \frac{w}{L} \quad \text{and} \quad \frac{d}{L} \quad , \tag{4.6}$$

constructed so that d , which we wish to isolate, occurs in only one of them. We could now proceed (in the absence of a satisfactory algorithm for continuous approximation) by computing a set of "data values" of these three parameters by means of (4.2), and then, as in the problem of Section 2, fitting a bivariate polynomial to these values. However, the practical requirement was that the formula derived should be as simple as possible, and the above approach is unlikely to achieve that. Clearly, there are different ways in which we can combine the five parameters to define our three variables, and some choices will lead to simpler formulae than others. A main task, therefore, will be to identify the most promising choices, and this, at least initially, can be done graphically. We shall be interested particularly in those choices which lead to relationships, defined through equation (4.2), that change most slowly over the parameter ranges.

For the latter reason, one choice we shall consider, instead of (4.6), is

$$x_1 = \left(\frac{F}{fL^2}\right)^{1/3} , \quad y_1 = \frac{w}{L} , \quad z_1 = \frac{d}{L} . \tag{4.7}$$

By introducing the cube root, as can most readily be seen from (4.1), we can expect the relationship to consist mainly of a linear, indeed a proportionate, relationship between z_1 and x_1, at any rate for lower values of d. Of course, the same would be true if we used $\log z_1$ and $\log x_1$, or z_1^3 and x_1^3, or indeed any other function of z_1 and x_1. Moreover, there is another choice of variables with the same properties as (4.7), namely

$$x_2 = \left(\frac{F}{fw^2}\right)^{1/3} , \quad y_2 = \frac{L}{w} , \quad z_2 = \frac{d}{w} . \tag{4.8}$$

Of the various possibilities tried, in fact (4.7) and (4.8) proved the most promising, and their respective relationships are shown in Figures 3 and 4. The full lines span the ranges given in (4.4). The broken lines are extensions which it would be convenient to cover if this did not complicate the formula required. The slope discontinuities, due to (4.2) having two parts, can clearly be seen.

Initially, because of the modest accuracy requirement, we had hoped that a single equation might suffice for the complete relationship, despite the discontinuities. From this point of view, Figure 4 does not look promising, with some of the curves having a long straight section followed by a curved section departing appreciably from it. Figure 3, with its shorter straight sections, looks better. However a difference of 2%, our accuracy requirement, is too small to be seen at the left-hand end of the curves: we need a diagram in which such a difference is clearly visible everywhere. This can easily be achieved by plotting z_1/x_1 instead of z_1. However, we can go a little further. By examining algebraically the behaviour of the first part of (4.2) close to the origin, we can establish the major part of the variation with y_1 in the "left-hand" relationship, and remove it from our plot. In fact, we find that for small values of z_1 we have approximately

$$z_1 = x_1 g_1(y_1) \quad , \tag{4.9}$$

where

$$g_1(y_1) = 2/[\pi(\tfrac{1}{2}\pi - 1 + 1/y_1)]^{1/3} \quad .$$

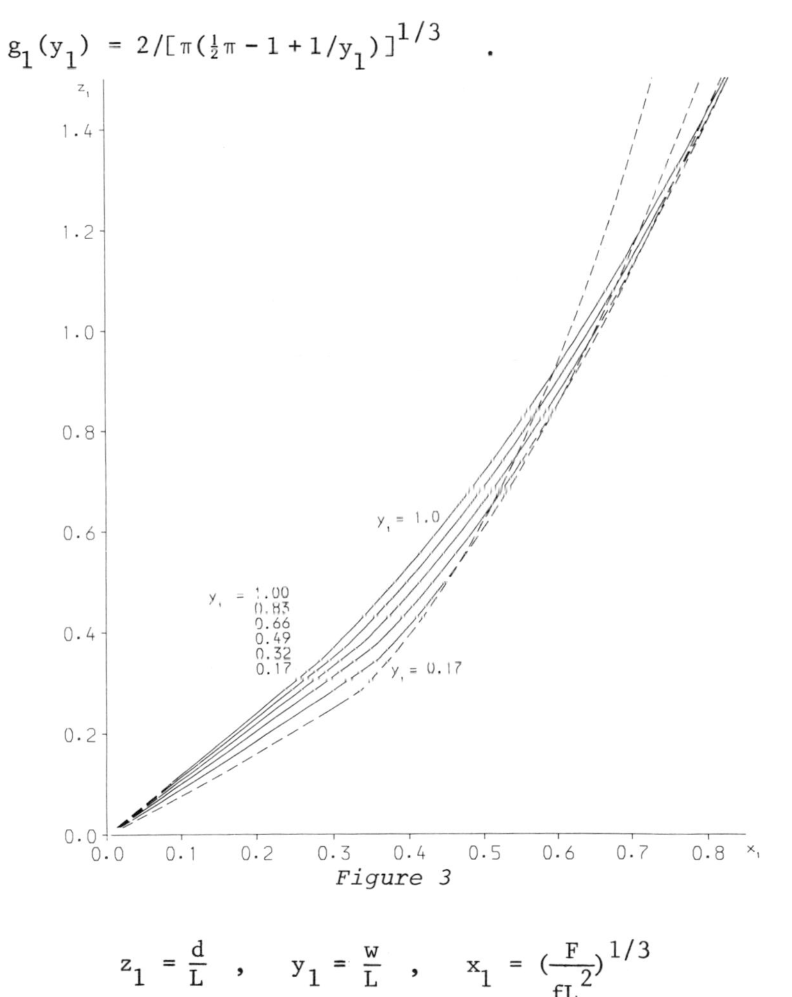

Figure 3

$$z_1 = \frac{d}{L} \quad , \quad y_1 = \frac{w}{L} \quad , \quad x_1 = \left(\frac{F}{fL^2}\right)^{1/3}$$

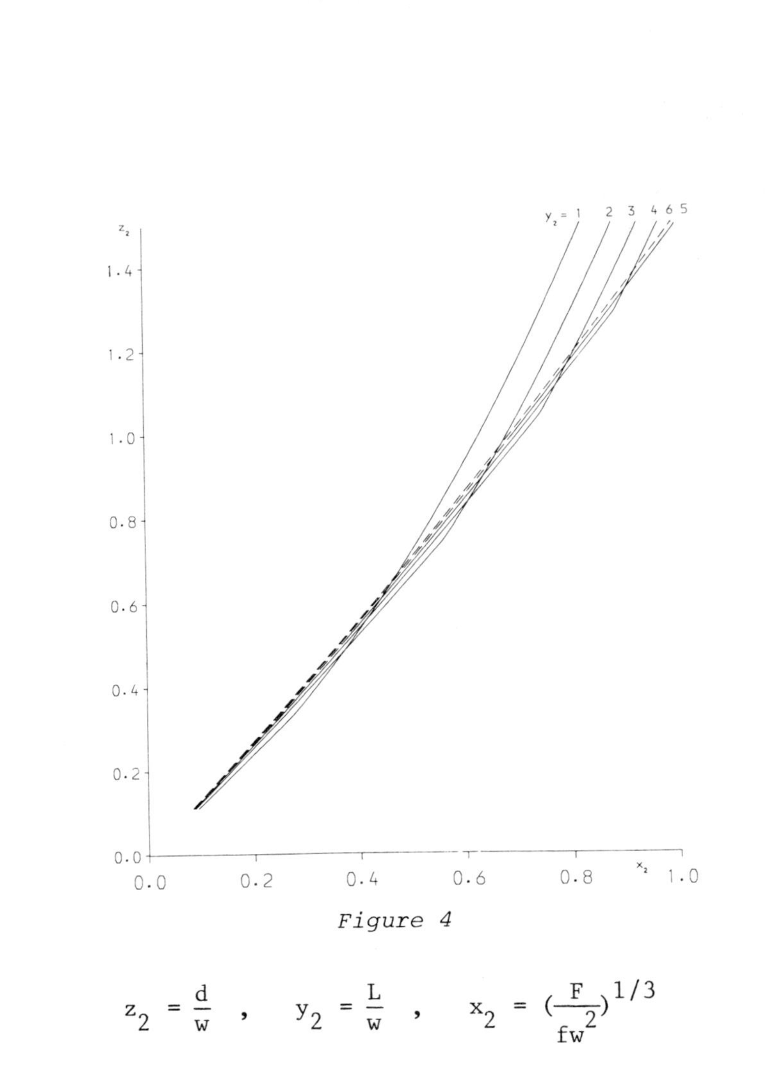

Figure 4

$$z_2 = \frac{d}{w} \quad , \quad y_2 = \frac{L}{w} \quad , \quad x_2 = \left(\frac{F}{fw^2}\right)^{1/3}$$

In Figure 5, therefore, we have plotted against x_1 the quantity

$$P = \frac{z_1}{x_1} - g_1(y_1) \quad . \tag{4.10}$$

Corresponding to an accuracy of 2% in z_1, the (absolute) accuracy required in P varies from about 0.015 to 0.035. It now becomes clear, however, that a simple smooth expression

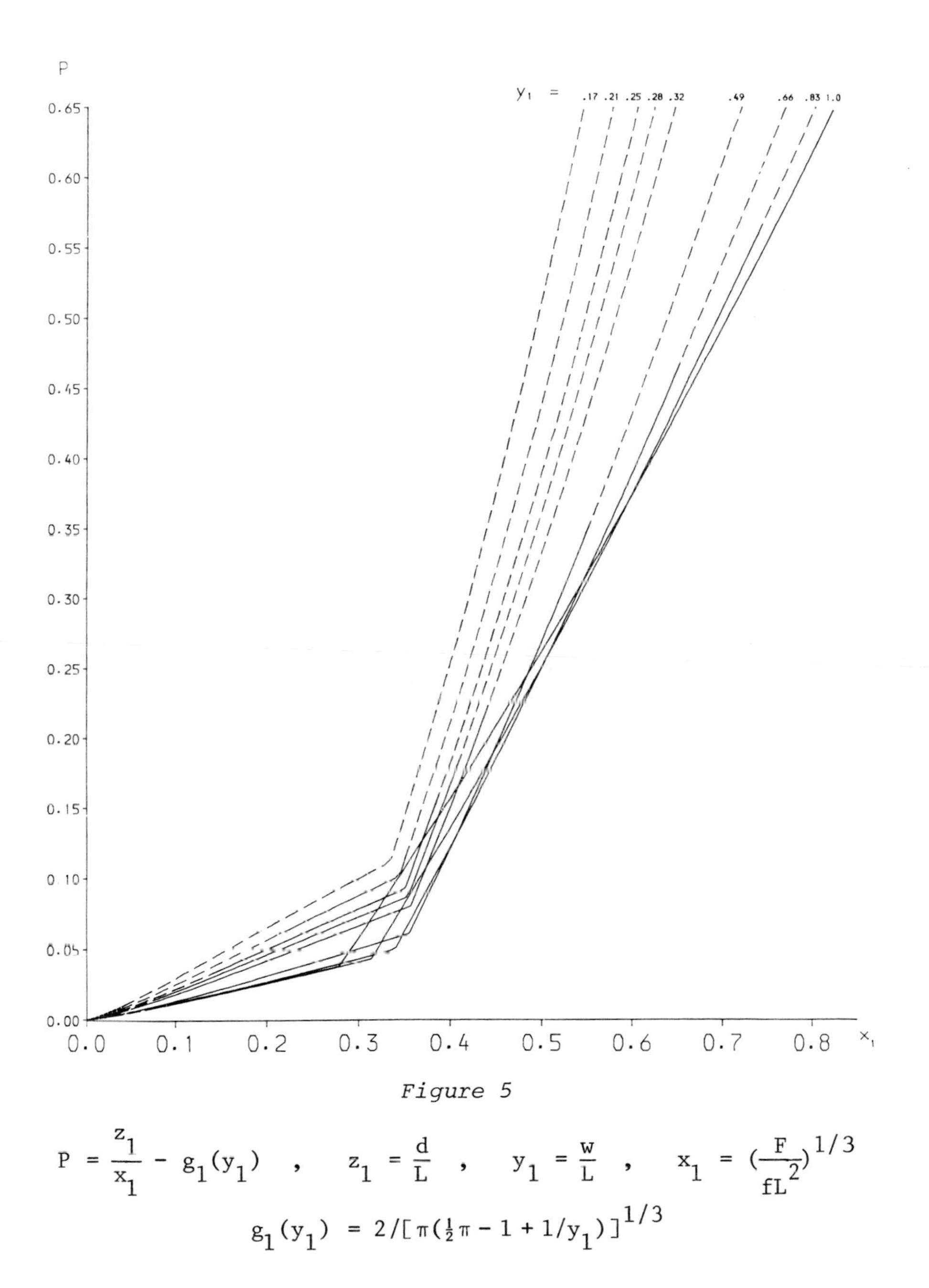

Figure 5

$$P = \frac{z_1}{x_1} - g_1(y_1) \quad , \quad z_1 = \frac{d}{L} \quad , \quad y_1 = \frac{w}{L} \quad , \quad x_1 = \left(\frac{F}{fL^2}\right)^{1/3}$$

$$g_1(y_1) = 2/[\pi(\tfrac{1}{2}\pi - 1 + 1/y_1)]^{1/3}$$

cannot approximate the full extent of these curves at all accurately. Thus we abandon our attempt to find a single formula and seek instead a pair of formulae, one for the first part of (4.2) and one for the second.

In this case, our reason for preferring the relationship in Figure 3 to that in Figure 4 disappears, and indeed we find that Figure 4 is now the more promising. The long straight segments on the left have become an advantage and their variation with y_2 is already quite small. Moreover, the right-hand segments are, to reasonable accuracy, not far from being parallel and equi-spaced, so that we can hope that a formula of the form

$$z = a_1 + b_1 x + c_1 x^2 + d_1 y \quad , \tag{4.11}$$

(in which the suffix 2 has been dropped from the variables) might be adequate, with an extra term or so if it is not. For the left-hand segments, again we need a clearer diagram, and so in Figure 6 we have plotted the quantity

$$Q = \frac{z}{x} - g(y) \quad , \tag{4.12}$$

where

$$g(y) = 2y^{1/3}/[\pi(\tfrac{1}{2}\pi - 1 + y)]^{1/3} \quad . \tag{4.13}$$

A 2% error in z now corresponds to an absolute error in Q of 0.025 to 0.030 . We can see that a single straight line will approximate all the segments to the left of the discontinuities (including their extensions) with an error little more than half this amount. Establishing that $g(y)$ can be accurately approximated by a straight line in $1/y$, we get

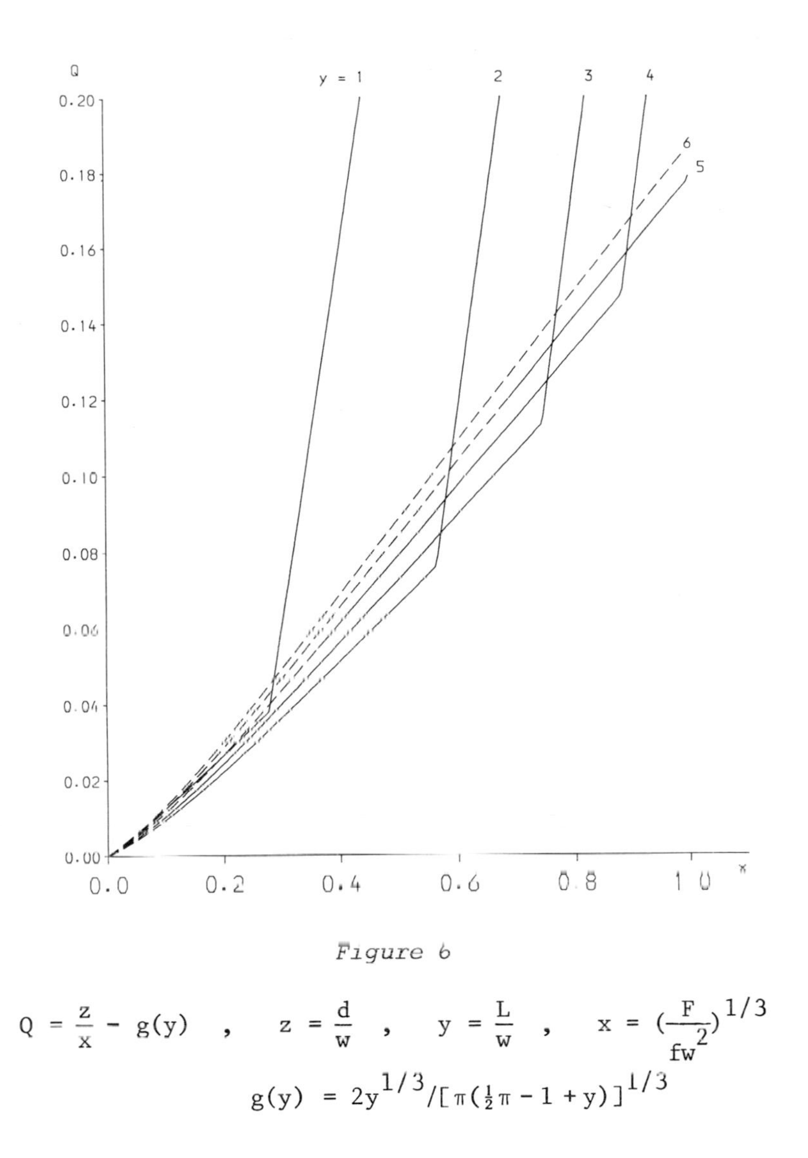

Figure 6

$$Q = \frac{z}{x} - g(y) \quad , \quad z = \frac{d}{w} \quad , \quad y = \frac{L}{w} \quad , \quad x = \left(\frac{F}{fw^2}\right)^{1/3}$$

$$g(y) = 2y^{1/3}/[\pi(\tfrac{1}{2}\pi - 1 + y)]^{1/3}$$

the trial form

$$z = x(a_2 + b_2 x + c_2/y) \quad . \qquad (4.14)$$

To obtain values for the coefficients, "data points" were computed from (4.2), and equation (4.14) was fitted to these using the Chebyshev norm appropriate to percentage errors. Care was taken in the choice of the points to ensure that the maximum percentage error was close to that of the continuous solution. The equation obtained was

$$z = x(1.332 + 0.188x - 0.179/y) \quad . \tag{4.15}$$

Fitting (4.11) similarly, we obtained

$$z = 0.097 + 1.019x + 0.977x^2 - 0.121y \quad . \tag{4.16}$$

The maximum errors were 1.3% and 1.6% respectively. Since these are well within requirements, we took advantage of the rather fortunate numerical values to obtain the satisfactorily simple design formula

$$\frac{d}{w} = 0.2x(6.7 + x - \frac{w}{L}) \quad , \quad \text{or} \quad 0.1 + x(1+x) - 0.12\frac{L}{w} \quad , \tag{4.17}$$

whichever is the larger.
Here, $x = (\frac{F}{fw^2})^{1/3}$, and the formula is valid in the ranges

$$1 \le \frac{L}{w} \le 6 \quad \text{and} \quad 0.1 \le \frac{d}{w} \le 1.5 \quad ,$$

which include those in (4.4). The maximum error is 2.4% .

Having completed this problem, we were asked to undertake an investigation which was similar except it had an extra variable. It concerned another component of lifting gear, the bow shackle (Figure 7). The extra variable comes from the width of the shackle's jaw, w . The problem was prosecuted in the same way as for the chain link, though now of course, instead of a number of sets of curves to examine, there are a number of sets of sets of curves. I shall not go into details

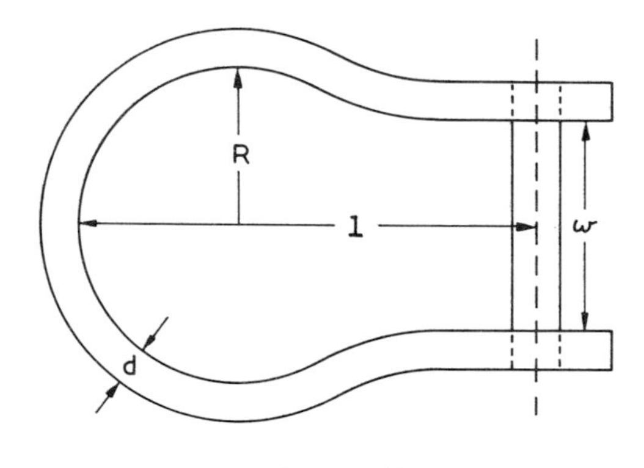

Figure 7

with this problem. It is mentioned mainly to indicate (a) that the process can be extended successfully to the case of three independent variables, but (b) that that it about as far as it can go: sets of sets of sets of curves, for the various possible pairings of variables, are altogether too cumbersome. The graphical process, in any case, is applicable only when data is provided (or, as in function approximation, can be obtained) at the nodes of a rectangular mesh in the appropriate number of dimensions.

We observe finally that, in the applications in this section, the computational algorithm played only a minor role: the investigation, dominated by the search for a good set of working variables and a suitable functional form, was essentially complete before the algorithm was brought into play.

5. DISTANT-WATER TRAWLERS (6 independent variables)

Our next application has six independent variables. It is described in Hayes[6] and concerns an attempt to express the hull resistance of a ship as a function of its shape, based on data available from towing-tank experiments with models of British distant-water trawlers. It is more typical of practical approximation problems than those of the previous section, in that the data points are haphazardly scattered in the space of the independent variables, and that the number of data points is severely limited. It is the latter feature which, in a case with several variables, prevents the fitting of a general polynomial in the manner of Section 2. The number of terms in the polynomial increases rapidly with the number of variables, whereas, in my experience, the number of data points provided does not. As a result there are usually insufficient data points to allow the fitting of a general polynomial of even quite modest degree, and it becomes a matter of trying to identify by some other means a set of terms which provides a good fit. This is true of the present application, with about 120 data points. (To illustrate how few this is, we remark that the number of corners of a 6-dimensional hypercube is 64.) Of course, we have to accept that the relationship concerned may be too complicated for the data points to describe it adequately. In that case, little can be done without extra data.

Another consequence of the larger number of variables and the scattered nature of the data is that it is not possible to seek a good choice of working variables by the graphical process of the last section. Indeed, because the search for a satisfactory fit with even one set of variables is likely to be a lengthy process, an investigation for a good choice of variables is not usually feasible by any numerical process. As a result, for such a choice one has to rely very much on the knowledge of the technologist in the subject, together with some general considerations which I shall discuss later. The six variables we used in our investigation to represent the shape of the hull were

L/B , B/d , C_m , C_p , LCB and α ,

where L , B and d are respectively the length, breadth and draft,

$$C_m = \frac{\text{area of maximum cross-section}}{B.d}$$

$$C_p = \frac{\text{volume}}{L \times \text{area of maximum cross-section}}$$

LCB = position of longitudinal centre of buoyancy,

and α measures the sharpness of the bow at a particular waterline.

All the above measurements relate only to the underwater portion of the hull. The dependent variable was, for technical reasons, not the hull resistance itself but a resistance coefficient C_R , corrected so as to relate to a common hull length and a common speed. Figure 8, with C_R plotted against displacement, indicates the amount of variation of C_R .

Initially, my naval architect colleague told me that there was no technical information available about the mathematical form of the relationship under investigation, so I first tried fitting a linear function of the six variables, hoping the residuals would provide some guidance. That, and a number of other early attempts, proved abortive. But then I began to realise, from discussions with my colleague, that there was after all some rather broad qualitative information available, so I decided to try modelling that information in the simplest mathematical terms. Thus, for example, it was known that, for fixed values of all the other variables, α must have a best value: too blunt an angle at the bow would increase resistance, and so would too sharp an angle, because it would cause sharp

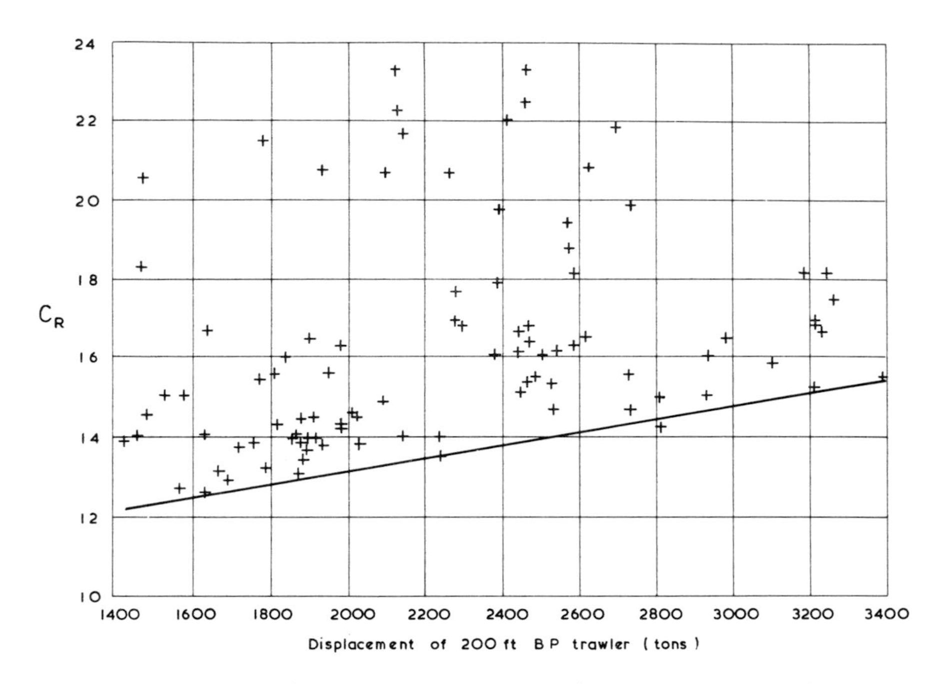

Figure 8 NPL Trawler Data for Conventional Forms. Design Speed (Trial), $V/\sqrt{L} = 1.10$

shoulders further back in the waterline. So there had to be at least a quadratic in α :

$$b_0 + b_1\alpha + b_2\alpha^2 \quad .$$

But the optimum α would be different for different values of C_p , so b_1 had to be at least a linear function of C_p . And at some values of C_p , α would be more critical than at others, in other words have a sharper optimum, so b_2 had to be at least linear in C_p . The same considerations applied with α and C_p interchanged in the argument, and so we get to the following set of terms, as being capable of providing the known type of behaviour

$$\begin{array}{lll} 1 & \alpha & \alpha^2 \\ C_p & \alpha C_p & \alpha^2 C_p \\ C_p^2 & \alpha C_p^2 & \end{array}$$

In fact, the corner term $\alpha^2 C_p^2$ was added, for no good reason other than it was tidier and provided a little fitting power to spare.

Sets of terms of this type then formed the basic components of the expression we built up. We considered each pair of parameters in turn, and the corresponding set of terms was added, or not added, according to my assessment of the reaction I got from my colleague. Ultimately, I assembled an expression of 32 terms, namely

$$\begin{aligned} C_R = {} & a_0 + a_1(B/d) + a_2(B/d)^2 + a_3(LCB) + a_4(LCB)^2 \\ & + a_5(C_p) + a_6(C_p)^2 + a_7(L/B) + a_8(L/B)^2 + a_9(C_m) \\ & + a_{10}(\alpha) + a_{11}(\alpha)^2 + a_{12}(C_p)(LCB) + a_{13}(C_p)(LCB)^2 \\ & + a_{14}(C_p)^2(LCB) + a_{15}(C_p)^2(LCB)^2 + a_{16}(C_p)(\alpha) \\ & + a_{17}(C_p)(\alpha)^2 + a_{18}(C_p)^2(\alpha) + a_{19}(C_p)^2(\alpha)^2 \\ & + a_{20}(C_p)(L/B) + a_{21}(C_p)(L/B)^2 + a_{22}(C_p)^2(L/B) \\ & + a_{23}(C_p)^2(L/B)^2 + a_{24}(L/B)(\alpha) + a_{25}(L/B)(\alpha)^2 \\ & + a_{26}(L/B)^2(\alpha) + a_{27}(L/B)^2(\alpha)^2 + a_{28}(B/d)(C_p) \\ & + a_{29}(B/d)^2(C_p) + a_{30}(B/d)(C_p)^2 + a_{31}(B/d)^2(C_p)^2 \end{aligned} \tag{5.1}$$

This form fitted the data to an accuracy almost as good as

the known experimental accuracy.

However, in a situation of this complexity, where we do not know how completely the data describes the situation, there are many pitfalls, and a good fit is not enough. It has to be proved in practice. In fact all turned out to be well with the derived equation and indeed we were able to use it to predict new designs with substantially lower hull resistance. Many trawlers were built to these new designs. The gains over previous best designs averaged about 15% , but sometimes were as high as 30% . Test results from models built to some of the new designs are shown in Figure 9.

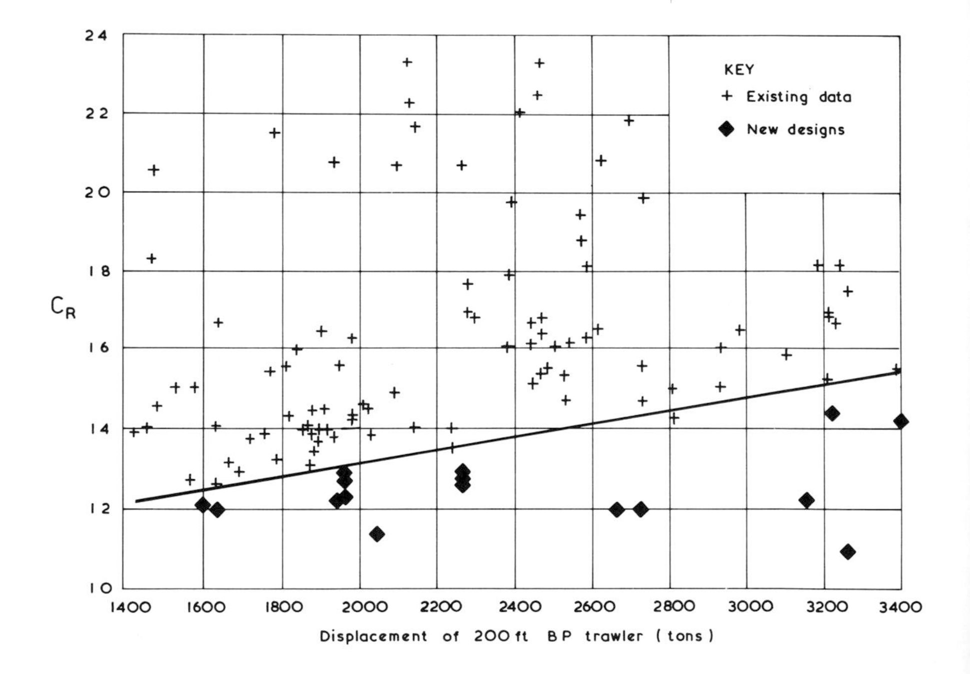

Figure 9 NPL Trawler Data for Conventional Forms. Design Speed (Trial), $V/\sqrt{L} = 1.10$

We now go back briefly to discuss the general considerations relating to the choice of variables for the fitting process. In the first place, the variables should be as independent as possible in their effect on the value of the dependent variable. The fewer the pairs of variables which interact, and the lower the degree of interaction of those which do interact, the smaller the number of terms which will be required in the mathematical model, and the more likely we are to be able to identify these terms. Secondly, with all other variables fixed, each variable should be capable of varying in a physically meaningful way over its full range, or at least a substantial part of it, regardless of the values of the other variables. If, instead, say two of the variables necessarily vary closely in step, it will not be possible, even if their effects are completely independent, for these effects to be satisfactorily distinguished from one another. The building of a mathematical model in this situation would be very difficult: we have seen how much depended on the technologist being able to envisage in a qualitative way the effect of varying each variable independently of the others, and how this effect altered with the value of a second variable. These points may be clearer if we consider an alternative set of variables suggested by another naval architect who considered our six variables insufficient to define the hull's shape. His set consisted of a number of offsets (the y values of Section 3) along each of a number of waterlines. Besides the much larger number of variables involved, which in itself is likely to prove prohibitive, it is fairly clear that each variable would intreact strongly with all its nearest neighbours, and probably with several of them simultaneously (with the set of variables actually used, it proved unnecessary, as we saw, to consider interactions between more than two at a time). Moreover, with all other offsets fixed, each offset could vary by only a small amount before it produced a bulge taking the hull outside the range of practical ship forms. In such forms, neighbouring offsets vary closely in step. The difficulties of such circumstances are, I hope, fairly clear.

6. SMALLER FISHING VESSELS (10/12 independent variables)

When news of the practical benefits stemming from the previous exercise had begun to spread, we were asked to undertake an even more daunting investigation in the same technical area. The request came from the Food and Agriculture Organization of the United Nations, who had collected, from various parts of the world, resistance results from towing tank tests related to smaller fishing vessels of particular relevance to developing countries. There were many more results available than there had been for the trawlers, in fact results on about 600 models, but that was an advantage, indeed a necessity. The extra difficulty came from two sources. The main one was the much greater variety in the design of the vessels, and therefore much greater ranges of the variables. (For example, the values of C_m, the ratio of the area of the largest (midship) cross-section to the area of its circumscribing rectangle, varied from 0.44 to 0.97. The former ratio, less than that for a simple triangular cross-section, could be that of a yacht, the latter could be that of a large oil-tanker. In the trawler work, C_m varied from 0.84 to 0.91.) The minor complication was the heterogeneity of experimental conditions at the various towing tanks, particularly the size of the tank itself (all the trawler models had been run in the same tank).

The first consequence of the greater variety of hull design was the need to introduce extra variables so as to describe that greater variety adequately. In fact, four extra variables were added for this purpose. Two more were added to take into account the differences between tanks (one of these, however, and one of the other ten, were simple zero-one variables). Then, starting with the same set of terms as used in the earlier investigation, we added further terms involving the new variables and their interactions with the old, chosen as before on the basis of technical experience, until we had about 50 terms. These proved inadequate, and it was at about this stage that we decided it would be wise for the time being to take a smaller bite at the problem. The data fell naturally into two parts: about half the data had come from Japanese tanks, the other half mainly from European tanks, with a few American results. We decided to work with just the latter half: the ranges of the variables, while still wide, were significantly smaller in this half than in the whole, and this might help us to make further progress.

The fifty terms were still inadequate, but this was no surprise. Technical inspiration, as to which pairs of variables to link together, had run out and moreover, because of the wider ranges, we had to expect that higher degree terms would be required, as well as products involving three, or perhaps more, variables. I therefore embarked on an extensive numerical investigation, examining the effect on the fit of adding different new groups of terms. Many were tried, and eventually a satisfactorily close fit was achieved. It contained 86 terms[2]. The fit still had to be tested in practice, and four new models were designed, built and tested for this purpose. Their results, together with the data, are plotted in Figure 10, taken from Traung, Doust and Hayes[10].

There remained the question of the Japanese data. The new form, with its 86 terms, was fitted to this segment of data

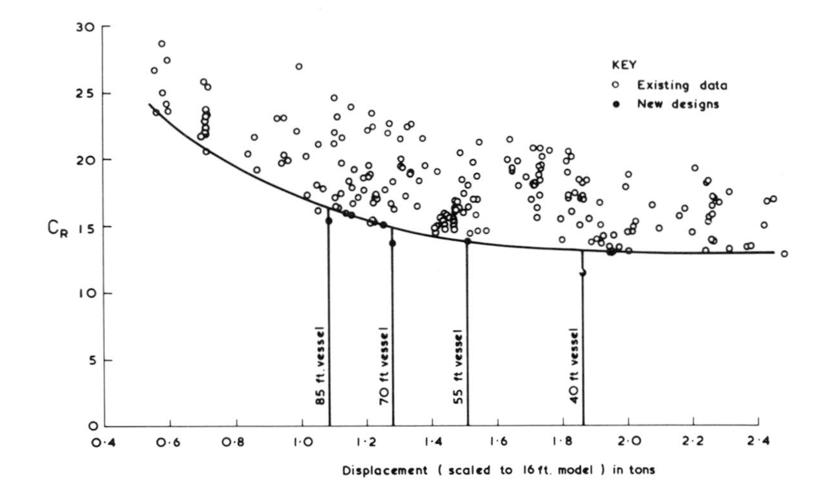

Figure 10 Results for the 40, 55, 70 and 85-footers at 1:1 speed length ratio compared with existing data

alone, but proved less satisfactory. So another numerical search for further terms was stated. There was little progress but then clues began to appear, in the changes in the fitted coefficients from one fit to the next, that some peculiarity in the data was causing the trouble. It was traced to a group of 42 vessels which in shape could perhaps best be described as outsize punts. They had not stood out, as incompatible data usually do, by producing unusually large residuals in the fits, because of their number and their unusual combination of parameter values. When these vessels were discarded, an 86-term expression very little different from the European form fitted the data satisfactorily. Moreover, the same expression proved satisfactory when fitted to the European and Japanese data combined. With this result established, the final stage was to go back and consider some of the terms in the derived expression which had not previously been investigated numerically. It was found possible to reduce the 86 terms to 72 (Hayes and Engvall[8]).

We see then that in the research described in these last two Sections, the design of a suitable form to fit to the data again constituted essentially the whole investigation. In one case, though, a new feature has appeared: the identification of incompatible data. In small problems this is not usually difficult, but in large problems, with many variables, such data can get lost in the complexity of the situation. Finally we note that again the numerical algorithm played a relatively minor role. Certainly it had very extensive use but it was not in any way a sophisticated algorithm. In fact, the computations, mostly completed before the stable method of Golub[4] became available, were carried out with an algorithm which followed the traditional route of the normal equations formed directly from powers and products of the variables (of course, each of the variables was normalized to the range -1 to $+1$). This proved quite satisfactory, despite the fact that the observation matrix was at times as large as 600×90.

7. CONCLUDING REMARKS

A number of multivariate applications has been presented, in 2 up to 12 independent variables, with a view to illustrating the fact that, as the numbers of dimensions increases, the importance of the numerical algorithm diminishes and other considerations become paramount. In the bivariate problem of Section 2, a good numerical algorithm was indeed the most important factor, making the choice of the polynomial degree a fairly straightforward matter. In the much more difficult bivariate problem of Section 3, the algorithm was again important, but so also was the choice of a suitable form for the approximating function. In the problems of Section 4, with 2 and 3 independent variables, design of the approximation basis constituted essentially the whole problem. We note also that, in these cases, the choice was mainly a matter of identifying, through numerical experiment, a good set of working variables. In the 6/12 variable problems of Sections 5 and 6, the choice of approximating function was again predominant. The larger number of dimensions prevented a numerical search for an advantageous set of variables, and to choose a set we relied to a major extent on the technologist's expert knowledge of the problem's context. His qualitative understanding was also important in the early stages in selecting an appropriate function of these variables, though later on extensive numerical experiment took over. The algorithm used was, however, quite primitive.

Such numerical experiments bear some resemblance to the step-wise search method of Efroymson, mentioned in the Introduction. To make physical sense in the non-linear situation, however, it is necessary to work with groups of terms rather than with individual terms, and to restrict the order in which the groups can be considered: it is not usually sensible, for example, to include higher degree terms in the fit if the corresponding lower degree terms are not already present. Moreover, it is advisable to monitor closely the coefficients obtained at each step of the search rather than to rely solely on a statistical significance test: we have seen for example how such a monitoring resulted in the identification of incompatible data, the consideration which proved crucial in the final stages of the investigation in Section 6.

In readily accepting my proposal to talk about applications rather than numerical algorithms, one of the organizers opined that theoreticians did not mind hearing about the real world as long as they didn't have to work in it. I hope that that is the case and that the problems presented have therefore been of some interest in themselves. This organizer went on to remark that, in any case, cross-fertilization was one of the objectives of the conference. Pursuing this rather agricultural analogy, perhaps what I have been trying to do is to apply the manure.

ACKNOWLEDGEMENT

The author is indebted to his colleagues M.G. Cox and D.W. Martin for their careful reading of the manuscript, resulting in many improvements to it.

REFERENCES

1. Anthony, G.T., Gorley, T.A.E. and Hayes, J.G. (1977). A New Design Formula for Chain Links, *NPL NAC Report* No.79.

2. Doust, D.J., Hayes, J.G. and Tsuchiya, T. (1967). A Statistical Analysis of FAO Resistance Data for Fishing Vessels, *in* "Fishing Boats of the World, 3" (Ed. J.-O. Traung) 123-138, London: Fishing News (Books) Ltd.

3. Efroymson, M.A. (1960). Multiple Regression Analysis, *in* "Mathematical Methods for Digital Computers" (Ed. A. Ralston and H.S. Wilf) 191-203, New York: John Wiley.

4. Golub, G.H. (1965). Numerical Methods for Solving Linear Least Squares Problems, *Numer. Math.* 206-216.

5. Hayes, J.G. (1962). The Mathematical Fairing of Ship Lines, *NPL Ship Report* No.42, 77 84.

6. Hayes, J.G. (1964). "The Optimum Hull Form Parameters", Proc. NATO Seminar on Numerical Methods Applied to Shipbuilding (Oslo-Bergen, September-October 1963, Oslo, Central Institute for Industrial Research, 295-323).

7. Hayes, J.G. (1970). Curve Fitting with Polynomials in One Variable, *in* "Numerical Approximation to Functions and Data" (Ed. J.G. Hayes) 43-64, London: Athlone Press.

8. Hayes, J.G. and Engvall, L.-O. (1969). Computer-Aided Studies of Fishing Boat Hull Resistance, *FAO Fisheries Technical Paper* No.87, Rome: Food and Agricultural Organization.

9. Hayes, J.G. and Halliday, J. (1974). The Least-Squares Fitting of Cubic Spline Surfaces to General Data Sets, *J. Inst. Math. Appl.* 14, 89-103.

10. Traung, J.-O., Doust, D.J. and Hayes, J.G. (1967). New Possibilities for Improvement in the Design of Fishing Vessels, *in* "Fishing Boats of the World, 3" (Ed. J.-O. Traung) 139-158, London: Fishing News (Books) Ltd.

MULTIVARIATE PIECEWISE POLYNOMIAL APPROXIMATION

John R. Rice

Mathematical Sciences
Purdue University
West Lafayette, Indiana, USA

1. INTRODUCTION

We begin with a brief historical development of convergence results for spline and piecewise polynomial approximation. This includes both univariate and multivariate material. Then we turn to the algorithmic question and discuss whether and, if so, how, one can actually achieve the dramatic power promised by the convergence results. In fact we can, which is in some contrast with classical polynomial approximation. We then point out, mostly by references, the difficulties with obtaining usable elements for multivariate approximation, especially if one desires to have smooth approximations.

We use the following general notation: $f(x)$ is a function to be approximated for $x \in D \subset R^N$. A partition of D is denoted by π and $|\pi|$ is the maximum volume of a piece of π. If S is a set of functions defined on D then $\mathrm{dist}_p(f,S)$ is the L_p distance of $f(x)$ from S, $\mathrm{dist}_{p,I}(f,S)$ is the same quantity with the domain restricted to a subset I of D. The volume of I is $|I|$. The function classes considered include $C^n(D)$, W_p^n (the Sobolev space) and Lip(α) (Lipshitz condition with exponent α) . The space of piecewise polynomials (including splines) of order n (degree $\leq$ n-1) on a partition π of D is denoted by S_π^n and $K = K(\pi)$ is the number of pieces in π . Splines are C^n elements of S_π^n.

2. CONVERGENCE RESULTS FOR PIECEWISE POLYNOMIAL APPROXIMATION

A. Univariate Results

The basic result is as follows:

THEOREM 1 *Let* π *be a partition of* [0,1] *with* K *pieces and* S^n *be the splines of order* n *defined on* π *. Then for* $j = 0,1,\ldots,k-1$ *there is a constant* C *depending only on* j *and* n *so that for all* $f(x) \in C^j[0,1]$ *we have*

$$dist_\infty(f, \text{splines } S_\pi^n) \leq C|\pi|^j w(f^{(j)}, |\pi|)$$

where $w(f,h)$ *is the modulus of continuity. In case* $f(x) \in C^n[0,1]$ *then*

$$dist_\infty(f, \text{splines } S_\pi^n) \leq C|\pi|^n \|f^{(n)}\| \quad .$$

This result has a long history of gradually improved versions until this best possible result was established; a higher degree of convergence than $|\pi|^n$ (for arbitrary π) implies that $f(x)$ is a polynomial. Significant steps in the development of this result were made by Birkhoff and de Boor[3], Meir and Sharma[23] and de Boor[5]. We have the simple

COROLLARY *If* π *is uniform and* $f(x) \in C^n[0,1]$ *then*

$$\text{dist}_\infty(f, \text{ splines } S_\pi^n) = O(K^{-n})$$

It is this version of linear (fixed knots) spline approximation convergence that can be extended to a much broader class of functions by non linear (variable knot) spline approximation. The following elementary result is established by the construction shown in Fig. 1.

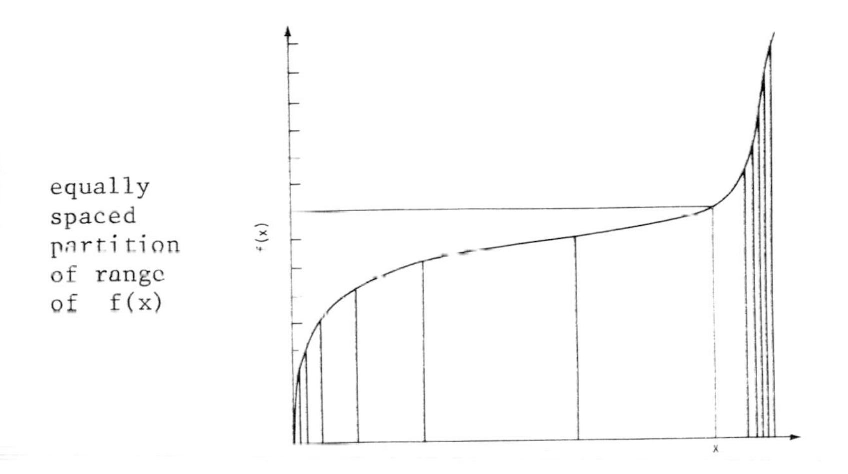

Figure 1 The construction for the proof of Theorem 2.

THEOREM 2 *Let* $f(x)$ *be continuous and monotone as* $[0,1]$ *with* $f(0) = 0$, $f(1) = 1$. *Then*

$$\text{dist}_\infty(f, \text{ step functions or broken lines}) \le \frac{1}{K}$$

A somewhat deeper result is that of Kahane[18].

THEOREM 3

$$\text{dist}_\infty(f, \text{ step functions}) = O(\frac{1}{K})$$

if and only if $f(x)$ *is continuous and of bounded variation.* This shows that nonlinear piecewise polynomial convergence results are quite different from classical approximation theory where the derivatives of $f(x)$ play a central role. However, this result is far too simple a case to allow one to confidently predict (or even conjecture) the general case.

The first general results appeared about 1970 as

THEOREM 4 (Rice, 1969) *Suppose* $f(x)$ *has singularities* $S = \{s_i\}_{i=1}^{R}$ *and set* $w(x) = \prod_{i=1}^{R}(x-s_i)$. *Assume* $f(x) \in C^n$ *except on* S , *is in* $\mathrm{Lip}(\alpha)$ *everywhere for* $\alpha > -\frac{1}{p}$ *and*

$$|f^{(n)}(x)| \leq \text{Const.}\,|w(x)|^{\alpha-n}$$

Then, with $K = K(\pi)$ *and* π *variable, we have*

$$\mathrm{dist}_p(f, \text{ splines } S_\pi^n) = \mathcal{O}(K^{-n})$$

THEOREM 5 (Burchard[12] and McClure[21]) *Set* $\sigma = \dfrac{1}{n+1/p}$ *and*

$$||f||_\sigma = [\int_0^1 |f^{(n)}|^\sigma]^{1/\sigma}$$

If $f(x) \in C^n$ *then*

$$\mathrm{dist}_p(f, \text{ splines } S_\pi^n) \leq \text{Const. } K^{-n}||f||_\sigma$$

McClure only covered the $p = 2$ case. Burchard's proof was obtained in 1969, but there were delays in its publication.

Theorem 5 suggests that the σ-norm $||f||_\sigma$ plays a key role in the theory; note that the functions in Rice's result have a finite σ-norm. Note also that $\sigma < 1$ except the simplest case previously covered by Kahane. It was clear at the time that neither of these results was best possible and it was easy to conjecture that $\|f\|_\sigma < \infty$ was necessary and sufficient for $O(K^{-n})$ convergence. The conjecture was false as Burchard constructed $f(x)$ with $\|f\|_\sigma < \infty$ which does not have $O(K^{-n})$ convergence. Note, however, that these results already cover all the "real world" functions likely to arise in applications.

Improved results were obtained over the next several years until the following definitive result.

THEOREM 6 (Peetre and Bergh[25], Brudnyi[10] and Burchard and Hale[14] *Let* P_N = *space of polynomials of order* n *and*

$$V_{1/n} = \{f(x) \mid \sum_{I\subset\pi} (\mathrm{dist}_{\infty,I}(f,P_N))^{1/n} \leq \mathrm{Const}_f \text{ for all } \pi\}$$

Assume $f(x)$ *is locally bounded on* $(-\infty,\infty)$ *and let* $V^0_{1/n}$ *be the closure in* $V_{1/n}$ *of functions with compact support. Then we have*

(Jackson-type) $f \in V^0_{1/n}$ *implies* $\mathrm{dist}_\infty(f,S^n) = O(K^{-n})$

(Bernstein-type) $f \in C^0$ *and* $\mathrm{dist}_\infty(f,S^n) = O(K^{-n})$ *implies* $f \in V^0_{1/n}$

Different authors formulate the definition of $V_{1/n}$ differently, but it is a generalized bounded variation class (note that $\mathrm{dist}_{\infty,I}(f,P_n)$ may be expressed in terms of nth differences of f on I). Slightly less definitive results are available for L_p- approximation (see Burchard[13], Brudnyi[11]) and Burchard has made a considerably study of the nature of the functions in these new and unusual spaces. The technique of de Boor to "pull apart knots" allows one to apply Theorem 6 to spline approximation.

B. Multivariate Results

The univariate theory was more concerned with piecewise polynomials with some smoothness (e.g. splines) as discontinuous piecewise polynomial results are of a classical nature. Discontinuous results should be classical for multivariate approximation (at least for uniform partitions), but the earliest explicit formulations of such results seems to be Morrey[24] and Birman and Solomyak[4]. The basic result is

THEOREM 7 *Let* $f(x) \in W_p^n(D)$ *with* D *the unit cube in* R^N. *Let* π *be a uniform partition of* D *by cubes with side* h. *Then*

$$\mathrm{dist}_p(f, S_\pi^n) = \mathcal{O}(h^n) = \mathcal{O}(K^{-n/N}) \quad .$$

It is not possible to "pull apart knots" in R^N, $N > 1$ and thus the extension of this result to smooth approximations was not straightforward. Schultz[31] obtained similar convergence rates for splines (again with a uniform partition of the unit cube) that are tensor products. His analysis depends essentially on the tensor product nature of the approximation and is valid for L_∞ and L_2. A more general result is

THEOREM 8 (de Boor and Fix[7]) *Let* D *be an open connected set in* R^N *for which there exists a continuous extension operator* $W_p^n(D) \to W_p^n(E)$ *where* E *is a rectangular polygon containing* D. *Let* π *be a rectangular partition of* D *and* j *be a multi-index* $(j_1, j_2, \ldots, j_N)$. *Then there is a linear projector* F_π *(the quasi-interpolant) onto splines* S_K^n *defined on* π *such that*

$$\left\| \frac{d^j(f-F_\pi f)}{\partial x^j} \right\|_\infty \le C_j\, K^{-n+1+|j|}\, W_{n-1}(f,|\pi|) \qquad |j| \le n-1$$

where

$$W_{n-1} = \max_{|\gamma|=n-1} w\left(\frac{d^\gamma f}{dx^\gamma}, |\pi|\right) \quad .$$

Brudnyi[10,11] has strengthened Theorem 7 to include nonlinear (variable partition) approximation and has obtained extensions to R^N of results similar to Theorem 6 and its L_p analogs.

It is not much of an exaggeration to summarize the univariate results by: *One can approximate any "realistic" function from* S_K^n *with the maximal rate of convergence.* The key questions are then: *Can this promise of efficient approximation be fulfilled in practice?* and *Can this be done for multivariate functions?* The answer to the first question is *yes* as seen in Ichida and Kiyono[16] and Rice[30]. The older nonlinear least squares spline approximation algorithm of de Boor and Rice[8], while not terribly efficient, showed the power of splines to approximate well any "reasonable" set of data. The answer for multivariate approximation is still unknown, but there are real complications and difficulties to be overcome. The approaches to overcoming these are examined in the next two sections and the final section discusses the recent result of de Boor and Rice[9] which suggests that there are some intrinsic limitations on our future capabilities for multivariate approximation, i.e. there are some ordinary, real world functions where one cannot obtain the hoped for maximal rate of convergence.

3. ELEMENTS FOR MULTIVARIATE APPROXIMATION

An *element* is a specific type of partition subdomain I and associated representation of the piecewise polynomials defined on I . In one variable all I are intervals, the properties of various representations are well understood and the whole matter revolves about computational efficiency, stability and convenience, see de Boor[6]. This is not so for multivariate approximation the moment one requests any kind of smoothness in the approximation, even simple continuity. There are two excellent and recent surveys of this area. Schumaker[33] and Bernhill[1], and one can also examine Barnhill and Risenfeld[2] for a more detailed examination of the ramifications of the choice of elements for various applications.

Tensor products of intervals and functions of one variable are mathematically attractive and widely used in practice. This approach has one, sometimes fatal, drawback: there are many domains which cannot, by any stretch of the mathematics, be considered as Cartesian product of intervals.

Simplicial partitions (i.e. triangles, tetrahedrons, etc.) are widely used, but they pose unexpected problems. For example, it is highly nontrivial to automatically triangulate in a reasonable way a general domain in the plane, even restricting oneself to "nice" domains, see Lawson[19]. Once a triangulation is made, it is nontrivial to organize the triangles for convenient use, i.e. to answer basic questions like: which triangle contains the point (.65, .813)? which triangles are neighbours to a given triangle? These things have been dealt with, but it complicates everything considerably even in two dimensions.

It took many years (and some faux pas) to settle the following question: Given a general triangulation π of the plane, is it possible to define a C^1 piecewise cubic on π

which has some degrees of freedom available to approximate f(x) (e.g. to interpolate f(x) at a few points in each triangle)? The answer is no, but it can be done for certain triangulations (just which ones is not known). If this simple question took so much time and effort it is possible that only slightly more complicated questions (C^2 quintics in 3-space on tetrahedrons?) may be completely intractable. Thus the practitioner of multivariate approximation must become familiar with the catalogue of useful elements that have been found and there is constant search to discover others. See Powell and Sabin[26]. For example, there is evidence that "natural" partitions of common surfaces in 3-space (e.g. airplanes) "require" elements that have 3, 4 and 5 sides. No one knows how to join a polynomial defined on a 5 sided subdomain smoothly with polynomials defined on neighbouring 3 or 4 sided subdomains.

Finally, one should be aware of the *blending functions* which interpolate N-1 dimensional data to approximate an N-variable function. If the interpolation is done by polynomials this gives piecewise polynomials only when $N = 1$. The blending may be done recursively (i.e. approximate the N-1 dimensional data by blending N-2 dimensional data) to produce actual piecewise polynomials provided the partition is rather regular. There are come complications that arise in this recursion, but they can be overcome. See Gordon[15] and Barnhill and Riesenfeld[2] for more information. It is not unlikely that successful multivariate smooth approximation schemes will be blending function methods or piecewise polynomial models of them.

4. ADAPTIVE COMPUTATION

An adaptive computation is one that dynamically adjusts its

strategy to take into account the behaviour of the particular problem to be solved, e.g. the behaviour of the function to be approximated. The first adaptive algorithm in this area was by McKeenman[22] for quadrature (L_1-approximation). A good understanding of such algorithms is seen in Lyness[20] and a general analysis, including convergence proofs, for the class of functions in Theorem 4 is given by Rice[28]. Rice[29] carried this analysis over to general L_p-approximation and implemented several variants (there are thousands of them) of the basic adaptive idea. Birman and Solomyak[4] independently introduced an adaptive partition algorithm but its power was not used in that paper. Subsequent work by Brudnyi uses their algorithm in an essential way to obtain tight convergence results.

We outline a simple case of adaptive quadrature. Suppose $f(x)$ is concave on $[0,1]$ and we use the trapezoidal rule (broken line approximation). Once a few lines are in they can be extended to the left and right to form a set of small triangles which enclose the curve of $f(x)$. The area of each triangle is an upper bound on the error for that particular interval whose triangle has the largest error. This is done repeatedly and, as the examples given in Fig. 2 show, the algorithm handles various singularities as efficiently as one can integrate $\sin(x)$ by classical (broken line) methods. This strategy is exactly that proposed by Birman and Solomyak and while it is clumsy for practical use, it is quite convenient for a theoretic analysis. Other equally effective strategies are convenient for actual computation.

There is, in principle, no difficulty in extending adaptive algorithms to multivariate approximation. However, like some other aspects of multivariate approximation, one can expect the practical complications to increase significantly.

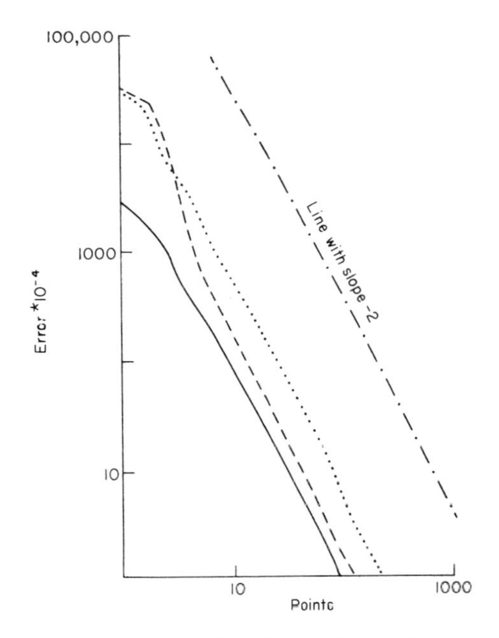

Figure 2 The convergence behaviour for the adaptive trapezoidal rule applied to $\sqrt[4]{x}$ (solid line), sin(x) (dotted line) and $f(x) = \min\ (7x-.5x^2, 5.28-30(x-.8)^{1.2})$ (dashed line). The convergence rate is $O(K^{-2})$ in each case.

5. AN ALGORITHM FOR MULTIVARIATE APPROXIMATION

We present here a new approach to the analysis of adaptive algorithms and outline the result recently obtained by de Boor and Rice[6] using it for multidimensional problems. The context considered is as follows:

Algorithm There is

1. A collection C of allowable cells.
2. A function $E: C \to R^1$ giving the error (bound) for approximation on a cell.
3. An initial partition of D into allowable cells.
4. A division algorithm for subdividing cells into two or more allowable cells.

An adaptive algorithm produces from the current partition a new partition by subdividing some cell. This continues until $E(C) \le \varepsilon$ for all cells C in the partition. Thus the prescribed ε specifies the accuracy of the final approximation. For the analysis we make the following assumptions.

Partition Cells

1. $\mathcal{C}$ consists of bounded, closed, convex sets.
2. For each $C \in \mathcal{C}$ there are balls $b_C \subseteq C \subseteq B_C$ so that

$$\eta = \inf \operatorname{diam}(b_C)/\operatorname{diam}(B_C) > 0 \quad .$$

3. If c = centre of b_C , then for all $\rho > 0$, $\mathcal{C}$ contains $C_p = c + \rho(C-c)$.
4. If the subdivision produces C^1 from C then $|C^1|/|C| \ge \beta > 0$.

ERROR If $C \le C^1$ then $E(C) \le E(C^1)$.

The new analytic tool is the function

$$\theta(x,\varepsilon) = \inf\{|C| \mid x \in C \in \mathcal{C} \text{ and } E(C) > \varepsilon\}$$

and the associated number

$$||E||_\varepsilon = \int_D \frac{dx}{\theta(x,\varepsilon)} \quad .$$

It may be shown that $\theta(x,\varepsilon)$ is bounded on bounded sets (unless $\equiv +\infty$) and is Lipshitz continuous on bounded sets. The principal result in this context is

THEOREM 9 *Let* π *be a partition produced by the adaptive algorithm from an initial partition* π_0 *where* $E(C) > \varepsilon$ *for all* $C \in \pi_0$ *. Then*

$$K(\pi) \leq \|E\|_\varepsilon / \beta \quad .$$

This allows one to bound the behaviour of the adaptive algorithm in terms of $\|E\|_\varepsilon$ and, in many cases, $\|E\|_\varepsilon$ may be bounded in terms of K as a function of ε .

This result is then applied to study the following approximation problem:

Domain D contains a smooth manifold S (of singularities) of dimension m .

Function f

1. $f^{(n)}(C) = \sup_{x \in C} \max_{|\gamma|=n} |\partial^\gamma f/\partial x^\gamma| \leq \text{const} * \text{dist}(S.C)^{\alpha-n}$
2. If $S \cap C \neq \phi$ then

$$\text{dist}_{p,C}(F,P_n) < \text{const} * |C|^{1/p} (\text{diam } C)^\alpha$$

(This approximation accuracy is attained by P_0 if $f \in \text{Lip}(\alpha)$) .

Error E

$$E(C) = \min\{F(C),G(C)\}$$

$$F(C) = \text{dist}(S,C)^{\alpha-n} (\text{diam } C)^n |C|^{1/p}$$

$$G(C) = (\text{dist } (S,C) + \text{diam } C)^\alpha |C|^{1/p}$$

THEOREM 10 *Suppose*

$$\alpha > m\, n/N - (N-m)/p \quad .$$

Then the algorithm as in Theorem 9 applied to this problem produces a partition π *such that*

$$K(\pi) \leq \text{const } \varepsilon^{-1/(n/N+1/p)}$$

or, equivalently,

$$\text{dist}_{p,D}(f, S_\pi^n) = O(K^{-n/N}) \quad .$$

In one variable there was no restriction on α except that f be in L_p, which, in R^N, merely requires $\alpha > -N/p$. However, de Boor and Rice show that the condition $\alpha > mn/N - (N-m)/p$ is, in general, *necessary* for the rate of convergence given. This approach has a wider application than piecewise polynomials and, for instance, applies to blending function methods.

Table 1 illustrates the smoothness requirements for the optimal convergence rates for L_1-approximation (adaptive quadrature).

Table 1 The smoothness requirements on f(x) for optimal order convergence, i.e. the number of continuous derivatives f(x) must have

Dimensions		Polynomial (Quadrature Rule) Order				
R	S	2	4	6	8	10
N=2	m=0,1					
	Point Singularities	-2	-2	-2	-2	-2
	Curve Singularities	0	1	2	3	4
N=3	m=0,1,2					
	Point Singularities	-3	-3	-3	-3	-3
	Curve Singularities	-4/3	-2/3	0	2/3	4/3
	Surface Singularities	1/3	5/3	3	4 1/3	5 2/3

6. ACKNOWLEDGEMENT

This work was supported in part by the National Science Foundation (GP 32940X) and the United States Army (DAAG-75-C-0024).

REFERENCES

1. Barnhill, R.E. (1977). Representation and Approximation of Surfaces, *in* "Mathematical Software III" (Ed. J. Rice) 69-120, Academic Press, New York.

2. Barnhill, R.E. and Riesenfeld, R.F. (1974). Computer Aided Geometric Design, Academic Press, New York.

3. Birkhoff, G. and de Boor, C.W. (1965). Piecewise Polynomial Interpolation and Approximation, *in* "Approximation of Functions" (Ed. H. Garabedian) 164-190, Elsevier, Amsterdam.

4. Birman, M.S. and Solomyak, M.Z. (1967). Piecewise Polynomial Approximations of Functions of Classes W_p^α, *Mat. Sbornik* 73, 295-317, also *Math. USSR Sbornik* 2, 295-317.

5. de Boor, C.W. (1968). On Uniform Approximation by Splines, *J. Approx. Theory* 1, 219-235.

6. de Boor, C.W. A Practical Guide to Splines (in preparation).

7. de Boor, C.W. and Fix, G. (1973). Spline Approximation by Quasi-interpolants, *J. Approx. Theory* 7, 19-45.

8. de Boor, C.W. and Rice, J.R. (1968). "Least Squares Cubic Spline Approximation II - Variable Knots", CSD-TR 21, Computer Science, Purdue University, also IMSL routine ICUVKU.

9. de Boor, C.W. and Rice, J.R. An Adaptive Algorithm for Multivariate Approximation Giving Optimal Convergence Rates (in preparation).

10. Brudnyi, Ju.A. (1974). Spline Approximation and Functions of Bounded Variation, *Dokl. Akad. Nauk SSR* 215, 511-513, also *Soviet Math. Dokl.* 15, 518-521.

11. Brudnyi, Ju.A. (1976). Piecewise Polynomial Approximation, Embedding Theorem and Rational Approximation, *in* "Approximation Theory, Bonn 1976" (Eds. R. Schaback and K. Scherer), Lecture Notes Math. 556, 73-98, Springer, Heidelberg.

12. Burchard, H.G. (1974). Splines with Optimal Knots Are Better, *Applicable Anal.* 3, 309-319.

13. Burchard, H.G. (1977). On the Degree of Convergence of Piecewise Polynomial Approximation on Optimal Meshes II, *Trans. Amer. Math. Soc.*, to appear.

14. Burchard, H.G. and Hale, D.F. (1975). Piecewise Polynomial Approximation on Optimal Meshes, *J. Approx. Theory* 14, 128-147.

15. Gordon, W.J. (1971). Blending Function Methods of Bivariate and Multivariate Interpolation and Approximation, *SIAM J. Numer. Anal.* 8, 158-177.

16. Ichida, K. and Kiyono, T. (1977). Curve Fitting by a One-Pass Method with a Piecewise Cubic Polynomial, *ACM Trans. Math. Software* 3, 164-174.

17. Jerome, J.J. (1973). Topics in Multivariate Approximation Theory, *in* "Approximation Theory" (Ed. G. Lorentz) 151-198, Academic Press, New York.

18. Kahane, J.P. (1961). Teoria Constructiva de Funciones, *Cursos y Seminarios de Matematica* 5, Univ. de Buenos Aires.

19. Lawson, C.L. (1977). Software for C^1-Surface Interpolation, *in* "Mathematical Software III" (Ed. J. Rice) 161-194, Academic Press, New York.

20. Lyness, J.N. (1970). Algorithm 379 - SQUANK (Simpson Quadrature Used Adaptively - Noise Killed), *Comm. Assoc. Comp. Mach.* 13, 260-263.

21. McClure, D.E. (1970). Feature Selection for the Analysis of Line Patterns, *Tech. Rpt. Div. Appl. Math.*, Brown University.

22. McKeenman, W.M. (1962). Algorithm 145 - Adaptive Numerical Integration by Simpson's Rule, *Comm. Assoc. Comp. Mach.* 5, 604.

23. Meir, A. and Sharma, A. (1966). Degree of Approximation by Spline Interpolation, *J. Math. Mech.* 15, 759-767.

24. Morrey, C.B. (1966). Multiple Integrals in the Calculus of Variations, Springer, New York.

25. Peetre, J. and Bergh, J. (1974). On the Spaces $V_p (0<p\leq\infty)$, *Boll. Un. Ital.* (4), 10, 632-648.

26. Powell, M.J.D. and Sabin, M.A. (1977). Piecewise Quadratic Approximations on Triangles, *ACM Trans. Math. Software* 3, to appear.

27. Rice, J.R. (1969). On the Degree of Convergence of Non-linear Spline Approximation, *in* "Approximation Theory with Special Emphasis on Spline Functions" (Ed. I.J. Schoenberg) 349-365, Academic Press.

28. Rice, J.R. (1975). A Metalgorithm for Adaptive Quadrature, *J. Assoc. Comp. Mach.* 22, 61-82.

29. Rice, J.R. (1976). Adaptive Approximation, *J. Approx. Theory* 16, 329-337.

30. Rice, J.R. (1978). Algorithm XXX - ADAPT: Adaptive Smooth Curve Fitting, *ACM Trans. Math. Software* 4, to appear.

31. Schultz, M.H. (1969a). L^{∞}-multivariate Approximation Theory, *SIAM J. Num. Anal.* 6, 161-183.

32. Schultz, M.H. (1969b). L^{2}-multivariate Approximation Theory, *SIAM J. Num. Anal.* 6, 184-209.

33. Schumaker, L.L. (1976). Fitting Surfaces to Scattered Data, *in* "Approximation Theory III (Eds. G. Lorentz, C. Chui and L. Schumaker) 203-268, Academic Press, New York.

A BLENDING FUNCTION INTERPOLANT FOR TRIANGLES

John A. Gregory

Department of Mathematics, Brunel University
Uxbridge, England

1. INTRODUCTION

This paper describes an interpolation scheme which, for integer $N \geq 0$, interpolates a function, and its derivatives of order N and less, on the entire boundary ∂T of a triangle T. Such a scheme is known as a "blending function" interpolant since it is constructed from a combination or blend of simpler interpolation functions. Alternatively the terms "smooth" and "transfinite" interpolation are used to describe such a scheme.

Blending function interpolation on triangles is introduced in the paper of Barnhill, Birkhoff, and Gordon[1], where rational interpolation functions are derived from Boolean sum combinations of interpolation operators. However, the basic scheme described in Section 2 of this paper is simply that of a linear combination of interpolation operators. This has the advantage of producing simple and symmetric interpolation schemes, although of somewhat lower accuracy than Boolean sum based schemes. Two examples of appropriate interpolation operators for the basic scheme of Section 2 are described in Sections 3 and 4 respectively. The first example results in an interpolation scheme which involves rational functions, whilst the second has the advantage that it results in a scheme which involves only polynomials.

A fuller account of some of the contents of the present paper can be found in [4]. Also, an extensive survey and bibliography of blending function interpolation, with particular reference to the representation and approximation of surfaces, is given in Barnhill[2]. Finally, the theory described in this paper can be extended to interpolation on n-dimensional simplices[5].

2. BASIC BLENDING FUNCTION INTERPOLATION SCHEME

In this section we define the basic interpolation scheme and in subsequent sections give two particular examples. Since, for the particular examples, the interpolation scheme is preserved by an affine transformation between two triangles, it is convenient to define the interpolant on a standard triangle T with vertices at $V_1 = (1,0)$, $V_2 = (0,1)$, and $V_3 = (0,0)$ in the (x,y)-plane, see Figure 2.1.

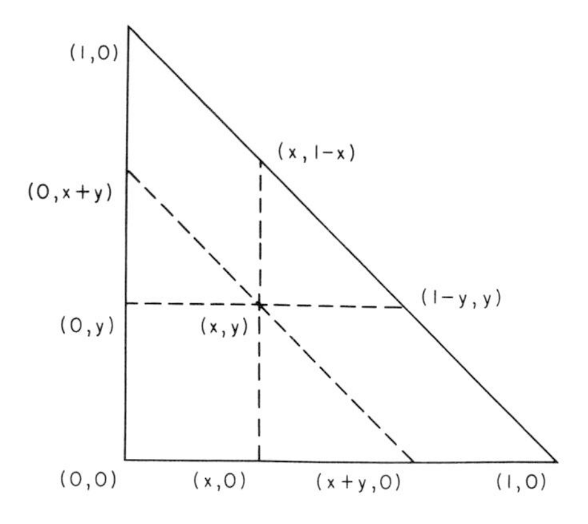

Figure 1

The boundary of T is denoted by ∂T and the side opposite the vertex V_k is denoted by E_k. Thus, E_1 is on the line $x=0$, E_2 is on the line $y=0$, and E_3 is on the line $z=0$, where $z \equiv 1-x-y$.

Let P_k; $k=1,2,3$, be linear operators and H be a subspace of $C^N(\bar{T})$ such that $P_k : H \to H$ and

$$D^{\nu}P_k[F]|_{E_i \cup E_j} = D^{\nu}F|_{E_i \cup E_j} \quad ; \quad i \neq j \neq k \neq i \quad , \tag{2.1}$$

for all $F \in H$ and $|\nu| = \nu_1 + \nu_2 \leq N$, where $\nu = (\nu_1, \nu_2)$ and $D^{\nu} = \partial^{\nu_1+\nu_2}/\partial x^{\nu_1}\partial y^{\nu_2}$. Thus, $P_k[F]$ is a function which interpolates F, and its derivatives of order N and less, on the sides of the triangle T excluding the side E_k. The interpolation scheme consists of a linear combination of the operators P_k and for this we introduce the following functions (the "blending functions"). Let

$$\begin{aligned}
\alpha_1(x,y) &= x^{N+1} \sum_{i=0}^{N} \sum_{j=0}^{N} y^i z^j (N+i+j)!/N!i!j! \quad , \\
\alpha_2(x,y) &= y^{N+1} \sum_{i=0}^{N} \sum_{j=0}^{N} z^i x^j (N+i+j)!/N!i!j! \quad , \\
\alpha_3(x,y) &= z^{N+1} \sum_{i=0}^{N} \sum_{j=0}^{N} x^i y^j (N+i+j)!/N!i!j. \quad ,
\end{aligned} \tag{2.2}$$

where $z \equiv 1-x-y$. Then

$$\alpha_1(x,y) + \alpha_2(x,y) + \alpha_3(x,y) = 1 \tag{2.3}$$

$$D^{\nu}\alpha_1|_{E_1} = D^{\nu}\alpha_2|_{E_2} = D^{\nu}\alpha_3|_{E_3} = 0 \quad ; \quad |\nu| \leq N \quad . \tag{2.4}$$

(These functions, and their properties, are derived from a Hermite type interpolant on the triangle T which for brevity is omitted here, see [4].)

The blending function scheme is now defined in the following theorem.

THEOREM 2.1 The linear operator $P : H \to H$ defined by

$$P \equiv \alpha_1 P_1 + \alpha_2 P_2 + \alpha_3 P_3 \tag{2.5}$$

is such that

$$D^{\nu} P[F]|_{\partial T} = D^{\nu} F|_{\partial T} \tag{2.6}$$

for all $F \in H$ and $|\nu| \le N$. That is $P[F]$ interpolates F, and its derivatives of order N and less, on the boundary ∂T of the triangle T.

Proof By symmetry it is sufficient to consider the side E_1 where, since $D^{\nu}\alpha_1|_{E_1} = 0$, it follows by Leibnitz' rule and the interpolation properties (2.1) that

$$\begin{aligned} D^{\nu} P[F]|_{E_1} &= D^{\nu}\{\alpha_2 P_2[F] + \alpha_3 P_3[F]\}|_{E_1} \\ &= \sum_{\mu \le \nu} \binom{\nu}{\mu} D^{\mu}(\alpha_2 + \alpha_3)|_{E_1} D^{\nu-\mu} F|_{E_1} \end{aligned}$$

for all $|\nu| \le N$. Now (2.3) and (2.4) give that

$$D^{\mu}(\alpha_2+\alpha_3)|_{E_1} = \begin{cases} 1 \; ; & |\mu| = 0 \;, \\ 0 \; ; & 1 \le |\mu| \le N \;, \end{cases}$$

and hence

$$D^{\nu} P[F]|_{E_1} = D^{\nu} F|_{E_1}$$

which completes the proof.

Remark 2.1 Linear interpolation operators which satisfy (2.1) are normally such that $P_k[G] \equiv 0$ for all $G \in H$ such that $D^{\nu}G|_{E_i \cup E_j} = 0$; $|\nu| \leq N$. This is certainly true for the particular examples given in the following sections. In such cases it follows that P_k ; $k = 1,2,3$, and P are projectors (i.e. idempotent linear operators) since then

$$P_k(I-P_k)[F] \equiv 0 \ ; \quad k = 1,2,3 \ , \quad \text{and}$$

$$P(I-P)[F] \equiv 0 \quad ,$$

where I is the identity operator.

Remark 2.2 The precision set of the operator P is defined as the set of all $F \in H$ such that $P[F] = F$. It follows, from (2.3) and (2.5), that if $P_k[F] = F$ for all $F \in \tau_k$; $k = 1,2,3$, then $P[F] = F$ for at least all $F \in \tau_1 \cap \tau_2 \cap \tau_3$. In particular, the schemes discussed in the following sections have the property that their precision sets contain $\mathcal{P}_{2N+1}$, where $\mathcal{P}_{2N+1}$ is the set of all polynomials of degree $2N+1$ or less. This polynomial precision set is important since it indicates the accuracy of the interpolation schemes.

3. RATIONAL HERMITE PROJECTOR SCHEME

Let

$$P_1[F](x,y) = \sum_{i=0}^{N} \phi_i\left(\frac{y}{1-x}\right)(1-x)^i F_{0,i}(x,0) + \sum_{i=0}^{N} \psi_i\left(\frac{y}{1-x}\right)(1-x)^i F_{0,i}(x,1-x) \quad , \tag{3.1}$$

$$P_2[F](x,y) = \sum_{i=0}^{N} \phi_i\left(\frac{x}{1-y}\right)(1-y)^i F_{i,0}(0,y) + \sum_{i=0}^{N} \psi_i\left(\frac{x}{1-y}\right)(1-y)^i F_{i,0}(1-y,y) \quad , \tag{3.2}$$

$$P_3[F](x,y) = \sum_{i=0}^{N} \phi_i\left(\frac{x}{x+y}\right)(x+y)^i \left(\left[\frac{\partial}{\partial x} - \frac{\partial}{\partial y}\right]^i F\right)(0,x+y) + \sum_{i=0}^{N} \psi_i\left(\frac{x}{x+y}\right)(x+y)^i \left(\left[\frac{\partial}{\partial x} - \frac{\partial}{\partial y}\right]^i F\right)(x+y,0) \quad , \tag{3.3}$$

where $\phi_i(t)$ and $\psi_i(t) = (-1)^i \phi_i(1-t)$ are the cardinal basis functions for Hermite two point Taylor interpolation on $0 \le t \le 1$. Thus

$$\phi_i^{(j)}(0) = \delta_{ij} \quad \text{and} \quad \phi_i^{(j)}(1) = 0 \ ; \quad 0 \le i,j \le N \quad , \tag{3.4}$$

where the $\phi_i(t)$ are polynomials of degree $2N+1$ defined by

$$\phi_i(t) = \sum_{k=i}^{N} (1-t)^{N+1} t^k \frac{(N+k-i)!}{N!i!(k-i)!} \quad . \tag{3.5}$$

Also, let H be the subspace of $C^N(\bar{T})$ of functions F which are $N+1$ times continuously differentiable along the sides of T, and whose derivative values which occur in the definitions of $P_k[F]$ above are N times continuously differentiable along the sides of T ; $k = 1,2,3$. Then the P_k are projectors on H which are appropriate for the interpolation scheme (2.5).

The projectors P_k are the Hermite two point Taylor projectors which act along parallels to the sides E_k between the sides E_i and E_j ; $i \ne j \ne k \ne i$, cf. Figure 2.1. These projectors, and hence the resulting interpolation scheme (2.5), are preserved by affine transformation. The fact that P : $H \to H$ is not obvious since the $P_k[F]$ involve rational terms which appear to have singularities at the vertices V_k. However, Barnhill and Gregory[3] show that $P_k[F] \in C^N(\bar{T})$ for all $F \in H$ and hence, from the interpolatory properties of $P_k[F]$, it follows that $P_k[F] \in H$ for all $F \in H$.

The precision set of the projector P includes the polynomial set $\mathcal{P}_{2N+1}$. This follows from Remark 2.2 since the intersection of the precision sets of the projectors P_k; $k = 1,2,3$, is the set of polynomials of degree $2N+1$ along parallels to the sides of T and this set includes $\mathcal{P}_{2N+1}$.

Example For illustrative purposes consider the case $N = 0$. Then

$$\begin{aligned} P[F](x,y) = {} & x\left[\frac{z}{1-x} F(x,0) + \frac{y}{1-x} F(x,1-x)\right] \\ & + y\left[\frac{z}{1-y} F(0,y) + \frac{x}{1-y} F(1-y,y)\right] \\ & + z\left[\frac{y}{1-z} F(0,x+y) + \frac{x}{1-z} F(x+y,0)\right] . \end{aligned} \tag{3.6}$$

4. POLYNOMIAL TAYLOR PROJECTOR SCHEME

Let

$$\begin{aligned} P_1[F](x,y) = {} & \sum_{i=0}^{N} (x+y-1)^i F_{i,0}(1-y,y)/i! \\ & + \sum_{j=0}^{N} y^j \left(\left[-\frac{\partial}{\partial x} + \frac{\partial}{\partial y}\right]^j F\right)(x+y,0)/j! \\ & - \sum_{i=0}^{N} \sum_{j=0}^{N} (x+y-1)^i y^j \left(\left[\frac{\partial}{\partial x}\right]^i \left[-\frac{\partial}{\partial x} + \frac{\partial}{\partial y}\right]^j F\right) \\ & (1,0)/i!j! , \end{aligned} \tag{4.1}$$

$$P_2[F](x,y) = \sum_{i=0}^{N} (x+y-1)^i F_{0,i}(x,1-x)/i!$$
$$+ \sum_{i=0}^{N} x^j \left(\left[\frac{\partial}{\partial x} - \frac{\partial}{\partial y} \right]^j F \right) (0,x+y)/j!$$
$$- \sum_{i=0}^{N} \sum_{j=0}^{N} (x+y-1)^i x^j \left(\left[\frac{\partial}{\partial y} \right]^i \left[\frac{\partial}{\partial x} - \frac{\partial}{\partial y} \right]^j F \right)$$
$$(0,1)/i!j! \quad , \qquad (4.2)$$

$$P_3[F](x,y) = \sum_{i=0}^{N} x^i F_{i,0}(0,y)/i! + \sum_{j=0}^{N} y^j F_{0,j}(x,0)/j!$$
$$- \sum_{i=0}^{N} \sum_{j=0}^{N} x^i y^j F_{i,j}(0,0)/i!j! \quad . \qquad (4.3)$$

Also, let H be the subspace of $C^N(\bar{T})$ of functions F whose derivative values which occur in the definitions of $P_k[F]$ above are N times continuously differentiable along the sides of T; $k = 1,2,3$, and which are such that

$$\left(\frac{\partial^{m+n} F}{\partial s_i^m \partial s_j^n} \right) (V_k) = \left(\frac{\partial^{n+m} F}{\partial s_j^n \partial s_i^m} \right) (V_k) \quad ; \quad 0 \le m,n \le N \; ;$$
$$k = 1,2,3 \; ; \; i \ne j \ne k \ne i \quad , \qquad (4.4)$$

where $\partial/\partial s_\ell$ denotes differentiation along the side E_ℓ. Then the P_k are appropriate projectors on H for the interpolation scheme (2.5).

The projectors P_k are derived from Boolean sum combinations of Taylor projectors which act on a side of the triangle T along parallels to another side, cf. Figure 2.1. The compatibility conditions (4.4) are necessary in order that $P_k[F]$ interpolate F and its derivatives both on E_i and E_j; $i \ne j \ne k \ne i$, and, in fact, ensure that the Taylor projector components of the $P_k[F]$ commute.

The polynomial precision set of the projector P_k ; $k = 1,2,3$, is

$$\xi_i^m \xi_j^n \begin{cases} 0 \le m \le N & \text{for all } n , \\ 0 \le n \le N & \text{for all } m , \end{cases} \qquad i \ne j \ne k \ne i , \tag{4.5}$$

where $\xi_1 = x$, $\xi_2 = y$, and $\xi_3 = z$. The intersection of these sets is the set $\mathcal{P}_{2N+1}$ which is thus contained in the precision set of the projector P , see Remark 2.2.

Example For illustrative purposes consider the case $N = 0$. Then

$$\begin{aligned} P[F](x,y) = {} & x[F(1-y,y) + F(x+y,0) - F(1,0)] \\ & + y[F(x,1-x) + F(0,x+y) - F(0,1)] \\ & + z[F(0,y) + F(x,0) - F(0,0)] \ . \end{aligned} \tag{4.6}$$

REFERENCES

1. Barnhill, R.E., Birkhoff, G. and Gordon, W.J. (1973). Smooth Interpolation in Triangles, *J. Approx. Theory* 8, 114-128.
2. Barnhill, R.E. (1977). "Representation and Approximation of Surfaces", MRC Mathematical Software Symposium, Academic Press, to appear.
3. Barnhill, R.E. and Gregory, J.A. (1975). Compatible Smooth Interpolation in Triangles, *J. Approx. Theory* 15, 214-225.
4. Gregory, J.A. (1974). "Symmetric Smooth Interpolation on Triangles, TR/34", Department of Mathematics, Brunel University.
5. Gregory, J.A. To be published 1979.

THE REPRESENTATION AND HANDLING OF SMOOTH THREE DIMENSIONAL SURFACES

K.R. Butterfield

*British Leyland, Oxford**

1. INTRODUCTION

Algorithmic considerations for the representation and handling of three dimensional curves and surfaces in terms of parametric splines are given. The representation of the spline in terms of B-splines has much to recommend it (Butterfield[2]; Cox[4]; de Boor[5]). Recurrence relationships between B-splines of consecutive orders are given, and the best of $(k-1)!/((k-1-m)!m!)$ schemes for the evaluation of the mth derivative of the B-spline basis stated. Such an algorithm is required for the construction of the linear system for the determination of the B-spline coefficients for the parametric interpolation and in the evaluation of the derivatives of the spline. The generalisation of the parametric curve interpolation to three dimensional surfaces is *not* straightforward. The modification of one of Coons'[3] surface patch representations presented permits parametric approximations of the required form and is straightforward to differentiate.

Consider the surface contours given by one of the following

---

*Now at Atkins Research and Development, Ashley Road, Epsom, Surrey.

(i) the intersection of a plane and a surface,

(ii) the intersection of two surfaces,

(iii) the projection of a curve onto a surface.

In addition to being able to compute points on a surface contour it is required to space those points subject to some *approximate* minimax criteria. Algorithms for such computations are also outlined.

2. THE B-SPLINE

When interpolating to smooth three dimensional data, parametric interpolation in terms of splines is effective. For the numerical interpolation, the representation of the spline in terms of B-splines has much to recommend it, from numerical stability and computational efficiency view points. It is the purpose of this section to show how the B-spline computations may be performed.

Let n and k be prescribed positive integers and $\pi = \{x_i\}_1^{n+k}$ be a k *extended partition* of the real line, that is $x_1 \le x_2 \le \ldots \le x_{n+k}$ with $x_j < x_{j+k}$ for all j. If, for $i = 1,2,\ldots,n$, $N_{i,k}(x)$ denotes the *normalised B-spline* of order k (degree $k-1$) defined upon the knots $\{x_j\}_{j=i}^{i+k}$ then $\{N_{i,k}(x)\}_{i=1}^{n}$ is a basis for polynomial splines of order k on π when $x_k^+ \le x \le x_{n+1}^-$. Since

$$N_{ik}(x) = (x_{i+k}-x_i)\, g_k(x_i, x_{i+1}, \ldots, x_{i+k}; x)$$

where

$$g_k(t;x) = \begin{cases} (t-x)^{k-1} & t \le x \\ 0 & t < x \end{cases}$$

the following recurrence relationships are readily derived, if $m \le k-2$

$$N_{j,k}^{(m)}(x) = \frac{k-1}{k-m-1}\left[\frac{x-x_j}{x_{j+k-1}-x_j} N_{j,k-1}^{(m)}(x) + \frac{x_{j+k}-x}{x_{j+k}-x_{j+1}} N_{j+1,k-1}^{(m)}(x)\right] , \qquad (2.1)$$

and for all m

$$N_{j,k}^{(m)}(x) = (k-1)\left[\frac{N_{j,k-1}^{(m-1)}(x)}{x_{j+k-1}-x_j} - \frac{N_{j+1,k-1}^{(m-1)}(x)}{x_{j+k}-x_{j+1}}\right] , \qquad (2.2)$$

where $N_{j,k}^{(m)}(x) = \int_{-\infty}^{x} N_{j,k}^{(m+1)}(s)ds$ when $m < 0$, and (i) if $x = x_r$, $x = x_r^-$ or $x = x_r^+$ is used throughout, (ii) if $x_{j+k-1}-x_j = 0$ or $x_{j+k}-x_{j+1} = 0$ factors involving such terms are considered zero. For a proof of (2.1) see Butterfield[2] and for (2.2) de Boor[5].

The following are assumed throughout: $x_i < x < x_{i+1}$, $0 \le m < k-1$, n_r are either 0 or 1 (when 1 differentiation of an associated matrix is implied) and $m = \sum_{r=1}^{k-1} n_r$. Let $\underset{\sim}{N}_{i,k}$, $E_{i,k}$ and $D_{i,k}$ be $k \times 1$, $(k+1) \times k$ and $k \times k$ matrices respectively where

$$\underset{\sim}{N}_{i,k} = \{N_{i-k+1,k}(x), \ldots, N_{i,k}(x)\}^t ,$$

$$E_{i,k} = \begin{bmatrix} x_{i+1}-x & & & \\ x-x_{i-k+1} & x_{i+2}-x & & \\ & \cdots & & \\ & & x-x_{i-1} & x_{i+k}-x \\ & & & x-x_i \end{bmatrix}$$

and

$$D_{i,k} = \begin{bmatrix} \frac{1}{x_{i+1}-x_{i-k+1}} & & \\ & \cdot & \\ & & \frac{1}{x_{i+k}-x_i} \end{bmatrix}$$

Note for $x_i < x < x_{i+1}$ the non-zero B-splines are included in $\underset{\sim}{N}_{i,k}$. It follows from (2.1) and (2.2) that the mth derivative of $\underset{\sim}{N}_{i,k}$ denoted by $\underset{\sim}{N}_{i,k}^{(m)}$ may be written

$$\underset{\sim}{N}_{i,k}^{(m)} = \frac{(k-1)!}{(k-m-1)!} E_{i,k-1}^{(n_{k-1})} D_{i,k-1} \ldots E_{i,1}^{(n_1)} D_{i,1} \quad . \tag{2.3}$$

Note since n_r is 0 or 1 the number of ways of choosing the m non zero n_r from (k-1) possibilities is $(k-1)!/((k-m-1)!m!)$. It is now seen that $\underset{\sim}{N}_{i,k}^{(m)}$ may be computed from the recurrence relationship

$$\underset{\sim}{v}^1 = (1) \;, \quad \underset{\sim}{v}^j = E_{i,j-1}^{(n_{j-1})} D_{i,j-1} \underset{\sim}{v}^{j-1} \quad , \qquad j = 2,\ldots,k \;,$$

where $\underset{\sim}{v}^j$ is a $j \times 1$ vector. In Butterfield[2] a detailed running error analysis (Peters and Wilkinson[14]) of the $(k-1)!/((k-m-1)!m!)$ possible factorisations was conducted. Let us denote by scheme A the choice $n_1 = n_2 = \ldots = n_{k-m-1} = 0$, $n_{k-m} = \ldots = n_{k-1} = 1$. There are distinct advantages in the use of scheme A for the computation of the mth derivative of the B-spline basis from numerical stability and computational efficiency considerations, the number of floating point arithmetic operations for scheme A being

		m = 0	1	2	3	k = 4
/	$\min(k-1-m,1)+\frac{1}{2}k(k-1)$	7	7	7	6	
*	$(k+1-m)\max(k-m-2,0)$	10	4	0	0	
-	$2(k-1-m)+(k+1)^2$	15	13	11	9	

The interpolation to smooth three dimensional data is now considered. Let $\underset{\sim}{S}(x)$ denote the spline,

$$\underset{\sim}{S}(x) = \sum_{j=1}^{n} \underset{\sim}{b}_j N_{j,k}(x) \tag{2.4}$$

where $\underset{\sim}{b}_j$ is a 3×1 vector. We now consider the data for the interpolation, that is $(t_i, d_i, \underset{\sim}{f}_i)$, $i = 1,2,\ldots,n$. Let

$$p_i = \begin{cases} t_i^+ & \text{if } d_i > 0 \\ t_i^- & \text{if } d_i < 0 \end{cases}, \quad i = 1,2,\ldots,n$$

where d_i is a non-zero integer. The p_i are values of x at which interpolation is required such that

$$x^+_{\max(i,k)} \leq p_i \leq x^-_{\min(i+k,n+1)} \tag{2.5}$$

and

$$\text{if } x_j^+ \leq p_i \leq x_{j+1}^- , \text{ then } x_j^+ \leq p_{i+1} \tag{2.6}$$

(Schoenberg and Whitney[15]). The $\underset{\sim}{f}_i$ are 3×1 vectors. The interpolation problem may now be stated as

B1: Determine $\underset{\sim}{S}(x)$ such that $\underset{\sim}{S}^{(|d_i|-1)}(p_i) = \underset{\sim}{f}_i$, $i = 1,2,\ldots,n$.

The equations defining B1 may now be written in matrix terms as $AB = F$ where A is a $n \times n$ stepped band matrix ((2.6) ensures a stepped band structure and (2.5) is a *necessary* condition for non singularity), B a $n \times 3$ matrix with $\text{row}_i B = \underset{\sim}{b}_i^t$ and F a $n \times 3$ matrix with $\text{row}_i F = \underset{\sim}{f}_i^t$. If r_i denotes the greatest row number in which, for the general form of A , a non-zero element occurs in the ith column then A may be stored compactly in $n \times k$ elements. It is then straightforward to solve the linear system efficiently and stably by a version of Gaussian elimination related to the stepped band structure with the upper bound $(n-1)(k-1)(k+3)+3nk$ on the number of long operations (multiplication or division). A suitable choice for the knot parameter x_j is most important, knot spacing related to the norm of the difference between consecutive data points is effective.

3. SURFACE REPRESENTATION

For curve representation knot spacing related to the norm of the difference between consecutive data points is effective. The practical generalisation of such parametrization to surfaces is non trivial. The surface interpolation given approximates such parametrization and gives 'satisfactory' results provided changes in the *rectangular topology* are not 'too extreme'.

Let $\underset{\sim}{r}_{ij}$, $i=1,2,\ldots,m$; $j=1,2,\ldots,n$, represent known three dimensional data forming a topological rectangular mesh. Consider the data to be interpolated by 4th order parametric splines whose nodes and knots correspond such that there are n splines where the jth interpolates $\underset{\sim}{r}_{1j},\ldots,\underset{\sim}{r}_{mj}$ and m splines where the ith interpolates $\underset{\sim}{r}_{i1},\ldots,\underset{\sim}{r}_{in}$. Let p and q be the parametrizations for these splines respectively, that is at the i,jth point the parametrizations p_{ij}, q_{ij} are such that $p_{i+1j}-p_{ij}=\|\underset{\sim}{r}_{i+1j}-\underset{\sim}{r}_{ij}\|$ and $q_{ij+1}-q_{ij}=\|\underset{\sim}{r}_{ij+1}-\underset{\sim}{r}_{ij}\|$. It is convenient to work in terms of patches.

Definition 3.1 The i,jth patch will be defined to be the surface bounded and continuous with the portion of the splines from $\underset{\sim}{r}_{ij}$ to $\underset{\sim}{r}_{i+1j}$, $\underset{\sim}{r}_{i+1j}$ to $\underset{\sim}{r}_{i+1j+1}$, $\underset{\sim}{r}_{i+1j+1}$ to $\underset{\sim}{r}_{ij+1}$ and $\underset{\sim}{r}_{ij+1}$ to $\underset{\sim}{r}_{ij}$.

Variables which increment by unity are now introduced.

Thus consider the ordered pair (u,v) such that $\underset{\sim}{r}_{ij}$ is associated with $(0,0)$, $\underset{\sim}{r}_{i+1j}$ with $(1,0)$ etc., non integer values being interpreted as lying within the patch. Let

$$[F_0\ F_1\ G_0\ G_1] = [u^3\ u^2\ u^1]\begin{bmatrix} 2 & -2 & 1 & 1 \\ -3 & 3 & -2 & -1 \\ 0 & 0 & 1 & 0 \\ 1 & 0 & 0 & 0 \end{bmatrix}$$

and define

$$g_{ij}(v) = F_0(v)(p_{i+1j}-p_{ij}) + F_1(v)(p_{i+1j+1}-p_{ij+1}) \quad ,$$

$$h_{ij}(u) = F_0(u)(q_{ij+1}-q_{ij}) + F_1(u)(q_{i+1j+1}-q_{i+1j}) \quad ,$$

$$p' = ug_{ij}(v) \quad \text{and} \quad q' = vh_{ij}(u) \quad .$$

When establishing continuity of first derivative between patches, the parametrization is assumed to be (p',v) or (u,q'). Thus

$$\frac{\partial}{\partial p'} = \frac{1}{g_{ij}(v)}\frac{\partial}{\partial u} \quad \text{and} \quad \frac{\partial}{\partial q'} = \frac{1}{h_{ij}(u)}\frac{\partial}{\partial v} \quad .$$

The surface representation $\underset{\sim}{\phi}_{ij}(u,v)$ in the i,jth patch is now defined by

$$\underset{\sim}{\phi}_{ij}(u,v) = \underset{\sim}{U}^t_{ij}(u,v)M_{ij}\underset{\sim}{V}_{ij}(u,v)$$

where

$$\underset{\sim}{U}^t_{ij}(u,v) = [F_0(u)F_1(u)g_{ij}(v)G_0(u)g_{ij}(v)G_1(u)] \quad ,$$

$$\underset{\sim}{V}^t_{ij}(u,v) = [F_0(v)F_1(v)h_{ij}(u)G_0(v)h_{ij}(u)G_1(v)] \quad ,$$

$$M_{ij} = \begin{bmatrix} A & A_{q'} \\ A_{p'} & A_{p'q'} \end{bmatrix}, \qquad A = \begin{bmatrix} \underset{\sim}{r}_{ij} & \underset{\sim}{r}_{ij+1} \\ \underset{\sim}{r}_{i+1j} & \underset{\sim}{r}_{i+1j+1} \end{bmatrix},$$

the suffices p' and q' denoting differentiation. The cross derivative term $A_{p'q'}$ may be evaluated by finite differences. It is straightforward to establish that we have continuity of position and first derivative across patch boundaries.

4. FOLLOWING SURFACE CONTOURS SUBJECT TO A MINIMAX CRITERIA

For the minimax approximation to surface contours, we may first approximate the surface contour by interpolation to surface contour points with a three dimensional parametric spline and then compute a minimax approximation for the spline. For the monimax approximation, planes may be constructed orthogonal to the spline. From the intersection of such planes with the surface contour new surface contour points may be obtained and then the spline fit, minimax computations repeated iteratively.

Consideration is now given to the minimax computation. A spline of Section 2. is approximated by a linear segment such that each end of the segment is a point on the spline and the maximum of the minimum distance from a point on the spline to the linear segment is a specified value. Thus let $H(x,u)$ denote the minimum distance from $\underset{\sim}{S}(x)$, $x_k^+ \leq x \leq u$ to the straight line from $\underset{\sim}{S}(x_k)$ to $\underset{\sim}{S}(u)$. If $\underset{\sim}{S}_x$ and $\underset{\sim}{S}_u$ denote $\underset{\sim}{S}(x)$ and $\underset{\sim}{S}(u)$ relative to $\underset{\sim}{S}(x_k)$ then

$$H(x,u) = ||PS_x||_2 \tag{4.1}$$

where P is the projector onto the subspace orthogonal to $\underset{\sim}{S}_u$, that is $P = I-\hat{\underset{\sim}{S}}_u\hat{\underset{\sim}{S}}_u^t$ where $\hat{\underset{\sim}{S}}_u = \underset{\sim}{S}_u/||\underset{\sim}{S}_u||_2$. The minimax problem may now be stated as:

M1 Determine ϕ_1, ϕ_2 such that when $x = \phi_1$, $u = \phi_2$

$$\text{(i)} \quad \Delta(u) = \max_{x_k^+ \le x \le u} H(x,u)$$

$$\text{and (ii)} \quad C(u) = \min_{x_k^+ \le u \le x_{n+1}^-} |\Delta(u) - \Delta| = 0$$

where Δ is a known constant.

If the contour is sufficiently linear then $C(\phi_2) \neq 0$, this may be detected as follows. Since $\underset{\sim}{N}_{i,k} \geq 0$ and $\|\underset{\sim}{N}_{i,k}\|_1 = 1$ it follows that $\|PS_x\|_p \leq \max_{i-k+1 \leq j \leq i} \|P(b_j - b_1)\|_p$, $p = 1,2,\infty$. Hence $C(\phi_2) \neq 0$ if when $u = x_{n+1}$, $\|P(b_j - b_1)\|_2 < \alpha\Delta$, $j = 1,2,\ldots,n$ where $\alpha \sim .5$.

Computationally a cross product representation of (4.1) would be used. For the plane intersections, (4.1) then simplifies and corresponds with that used by Marlow and Powell[13]. Indeed

$$H(x,u) = |G(x,u) = \sum_{i=1}^{n} c_i(u) N_{i,k}(x)|$$

where $c_i(u)$ are scalar functions and $G(x_k, u) = G(u,u) = 0$. The following Theorem and Corollary may now be used for the determination of $\Delta(u)$. From (2.3) the following are established.

THEOREM 4.1 A necessary condition for $\partial G/\partial x$ to have a zero x, $x_i < x < x_{i+1}$ is that $c_j(t) - c_{j-1}(t)$ and $c_{j-1}(t) - c_{j-2}(t)$ are of opposite sign, $i-k+3 \leq j \leq i$.

COROLLARY 4.1 For $k = 3$ the conditions of Theorem 4.1 are

sufficient and the root is given by

$$x-x_i = \frac{x_{i+1}-x_i}{x_{i+1}-x_{i-1}} \frac{c_{i-1}(t) - c_{i-2}(t)}{\dfrac{c_{i-1}(t)-c_{i-2}(t)}{x_{i+1}-x_{i-1}} - \dfrac{c_i(t)-c_{i-1}(t)}{x_{i+2}-x_i}}$$

Note if x_j and $c_j(t)$ are exact, then the right hand side of the above may be computed with a small relative error.

Consider the following problem:

$$\text{P1:} \quad \text{minimize} \quad \{F(x) = \underset{\sim}{f}^t \underset{\sim}{f}\} \quad , \quad x \in E^n$$

$$\text{subject to} \quad \underset{\sim}{l} \le \underset{\sim}{x} \le \underset{\sim}{u}$$

where $\underset{\sim}{f}$ is a $m \times 1$ vector of non linear functions. Algorithms for the solution of P1 may be used to determine points on the surface contours (i)-(iii) and also for the computation of the contour points corresponding to the spline minimax points. Let $\underset{\sim}{g}$ be the gradient of $F(x)$ and G the hessian matrix, then we have $\underset{\sim}{g} = 2J^t\underset{\sim}{f}$ where J is the Jacobian of $\underset{\sim}{f}$ $(J_{ij} = \partial f_i/\partial x_j)$, and $G = 2J^tJ + 2\sum_{k=1}^{m} f_kG_k$ where G_k is the hessian matrix of f_k $((G_k)_{ij} = \partial^2 f_k/\partial x_i \partial x_j)$.

Optimization algorithms are nearly all of the following form. Given $\underset{\sim}{x}^k$, the kth estimate of the solution, a new estimate is obtained by first determining a direction of search $\underset{\sim}{p}^k$ and then a step length α^k so that $\underset{\sim}{x}^{k+1} = \underset{\sim}{x}^k + \alpha^k \underset{\sim}{p}^k$. Since we will only be concerned with a single iteration the suffix k will be dropped.

For the univariate minimization to determine α, we may proceed iteratively by the use of a bracketing technique (Davidon's 1959 formula) with safeguards (Gill and Murray[9]). Let us consider the determination of $\underset{\sim}{p}$ neglecting the constraints. If the G_k terms are neglected then the orthogonal reduction of Businger and Golub[1] may be used and such an approach may have advantages from numerical stability considerations (Golub and Wilkinson[11]). However, from a singular value decomposition of J (Lawson and Hanson[12]) we see that for such an approach and also for the Marquardt modification (Fletcher[6]) $\underset{\sim}{p}^{(1)}$ the minimum length solution (J rank deficient) lies in the row space of J (Gill and Murray[10]). If the G_k terms are incorporated then the solution $\underset{\sim}{p}^{(2)} \in E^n$ and may be markedly different to $\underset{\sim}{p}^{(1)}$. From observations of numerical results $\underset{\sim}{p}^{(2)}$ is often to be preferred to $\underset{\sim}{p}^{(1)}$. Note that the Hessian G, which may be obtained by finite differences may not be positive definite.

For the numerical solution of P1 we may proceed in the spirit of Fletcher[1]. That is $\underset{\sim}{p}$ is the vector which minimizes $\frac{1}{2}\delta\underset{\sim}{x}^t C\delta\underset{\sim}{x} + \delta\underset{\sim}{x}^t\underset{\sim}{g}$, $\delta\underset{\sim}{x} \in E^n$, $||\delta x = \underset{\sim}{x} - x^k||_\infty \le h$, $\underset{\sim}{l} \le \underset{\sim}{x} \le \underset{\sim}{u}$ (Fletcher and Jackson[8]) where h is adjusted during each iteration according to the success of the search direction.

REFERENCES

1. Businger, P. and Golub, G.H. (1965). Linear Least Squares Solutions by Householder transformations, *Numerische Mathematik* 7, 269-276.

2. Butterfield, K.R. (1976). The Computation of All the Derivatives of a B-Spline Basis, *J. Inst. Maths. Applics.* 17, 15-25.

3. Coons, S.A. (1967). "Surfaces for Computer-Aided Design of Space Forms", Massachusetts Institute of Technology, MAC-TR-41.

4. Cox, M.G. (1972). The Numerical Evaluation of B-Splines, *J. Inst. Maths. Applics.* 10, 134-149.

5. De Boor, C. (1972). On Calculating with B-Splines, *J. Approximation Theory* 6, 50-62.

6. Fletcher, R. (1971). "A modified Marquardt Subroutine for Non-Linear Least Squares", Report R6799, UKAERE Harwell.

7. Fletcher, R. (1972). An Algorithm for Solving Linearly Constrained Optimization Problems, *Math. Prog.* 2, 133-165.

8. Fletcher, R. and Jackson, M.P. (1974). Minimization of a Quadratic Function of Many Variables Subject Only to Lower and Upper Bounds, *J. Inst. Maths. Applics.* 14, 159-174.

9. Gill, P.E. and Murray, W. (1974). "Safeguarded Steplength Algorithms for Optimization Using Descent Methods", Report NAC37 NPL.

10. Gill, P. and Murray, W. (1976). *In* "Numerical Analysis Dundee 1975" (Ed. G.A. Watson) Lecture Notes in Mathematics, Springer-Verlag.

11. Golub, G.H. and Wilkinson, J.H. (1966). Note on the Iterative Refinement of Least Squares Solution, *Numerische Mathematik* 9, 139-148.

12. Lawson, C.L. and Hanson, R.J. (1974). "Solving Least Squares Problems", Prentice-Hall Inc.

13. Marlow, S. and Powell, M.J.D. (1972). "A FORTRAN Subroutine for Drawing a Curve Through a Given Sequence of Data Points", AERE-R 7092.

14. Peters, G. and Wilkinson, J.H. (1971). Practical Problems Arising in the Solution of Polynomial Equations, *J. Inst. Maths. Applics.* 8, 16-35.

15. Schoenberg, I.J. and Whitney, Anne (1953). On Polyá Frequency Functions III. The Positivity of Translation Determinants with an Application to the Interpolation Problem by Spline Curves, *Trans. Amer. Math. Soc.* 74, 246-259.

A TECHNIQUE FOR APPROXIMATE CONFORMAL MAPPING

by

S.W. Ellacott

Department of Mathematics
Brighton Polytechnic
Brighton, Sussex

ABSTRACT

A method is presented for approximating the mapping of a Jordan region onto the unit disc. The approximation is based on Green's function and takes a particularly simple form. Some computed examples are given.

I. INTRODUCTION

We consider the problem of conformally mapping a Jordan region D bounded by a Jordan curve Γ onto the unit open disc S. Assume $0 \in D$. Let $\bar{D} = D \cup \Gamma$ denote the closure of D and $\bar{S}$ the closed disc. By the Riemann mapping theorem there exists a unique map ψ which satisfies $\psi(0) = 0$, $\psi'(0)$ real and positive. Once a mapping of D onto S has been determined, the mapping of D to other useful regions such as half planes or squares can be achieved using standard mappings. Existing methods for solving the problem outlined above are generally of two types: either integral equation methods[1,2] or methods in which the mapping is expanded in terms of some integral kernel[3]. For further references, see page 27 of [3]. Methods of the first type give only a numerical solution for the values of the mapping on the boundary:

interior points have to be determined by Cauchy's integral formula. The "Bergman Kernel Method" discussed by Levin, Papamichael and Sideridis in [3] gives the mapping function as an integral.

In this paper we discuss a method based on approximation theory which gives the approximate mapping function in a particularly simple form. Opfer (private communication) has pointed out that the problem can be expressed as a complex polynomial approximation problem, but mappings with singularities on or near the boundary cannot be efficiently expressed as polynomials. The method presented here uses real approximation and can be adapted to take account of such singularities. In the remainder of this introductory section we establish some notation and describe the technique. In later sections we justify and analyse the method, discuss some modifications and present some computed examples.

The basis of the method is the construction of an approximate Green's function for $\bar{D}$. $G(z) = \ln|\psi(z)|$ is called Green's function with respect to zero for the region. The following properties are well known.

(i) $G(z)$ is harmonic on $D - \{0\}$ and continuous on $\bar{D} - \{0\}$.

(ii) $G(z) = 0$, $z \in \Gamma$.

(iii) $\lim_{z \to 0} (G(z) - \ln|z|)$ exists and is finite.

(iv) $\psi(z) = \exp(G(z) + iH(z))$ where H is any harmonic conjugate of G . (The position of the branch cut that H must have is irrelevant, but H must be suitably chosen to make $\psi'(0)$ real and positive.)

We construct an approximate Green's function by approximating $\ln|g(z)|$ (in the Chebyshev sense) on Γ by the real part of a polynomial. Here $g(z)$ is any function analytic on D, continuous on $\bar{D}$ and which satisfies $g(0) = 0$, $g'(0) > 0$, $g(z) \neq 0$, $z \neq 0$. We can choose $g(z) \equiv z$ but a better choice may be possible by making use of available information about ψ.

If $p(z)$ is a polynomial of degree n, with $p(0)$ real, say

$$p(z) = a_0 + (a_1 - ib_1)z + \ldots + (a_n - ib_n)z^n$$

(a_j, b_k real, $j = 0,\ldots,n$, $k = 1,\ldots,n$) then

$$\mathrm{Re}(p(z)) = a_0 + \sum_{j=1}^{n} a_j \,\mathrm{Re}(z^j) + \sum_{j=1}^{n} b_j \,\mathrm{Im}(z^j)$$

which is a linear combination of elements of the set

$$\{1, \mathrm{Re}(z), \mathrm{Im}(z), \ldots, \mathrm{Re}(z^n), \mathrm{Im}(z^n)\} \quad .$$

Thus we can approximate $\ln|g(z)|$ using a standard technique for linear Chebyshev approximation, and we denote the polynomial corresponding to the best approximation by p_n.

The approximate mapping is then given by

$$\phi_n(z) = g(z) \exp(-p_n(z)) \quad .$$

II. ANALYSIS OF THE METHOD

THEOREM 1

The set of functions

$$\{1, \mathrm{Re}(z), \mathrm{Im}(z), \ldots, \mathrm{Re}(z^n), \mathrm{Im}(z^n)\}$$

is linearly independent (over the real numbers) on Γ .

Proof The result follows from the fact that a linear combination of the functions is harmonic, so if zero on Γ such a combination vanishes identically in $\bar{D}$.

The functions are not independent on an arbitrary curve, e.g. 1 , $\mathrm{Re}(z)$, $\mathrm{Im}(z)$ are linearly dependent on any straight line. Thus the basis will not in general form a Chebyshev set on Γ . Andreasson and Watson[5] have provided a recent survey of best linear approximation algorithms that do not require the Chebyshev set hypothesis. However for the examples given in the next section, Lawson's algorithm was used[6]. (See below.) Now let

$$\varepsilon_n = \| \ln|g(z)| - \mathrm{Re}(p_n(z)) \|$$

where $\| \ \|$ denotes the Chebyshev norm on Γ .

We have:

THEOREM 2

(a) $\lim_{n\to\infty} \varepsilon_n = 0$.

(b) Let $G(z)$ be Green's function (with respect to ∞) for the region outside Γ . For each $\mu > 0$ let $\Gamma_\mu = \{z|G(z) = \mu\}$ and $R = e^\mu$. Then if $\psi(z)/g(z)$ is analytic and non-zero on and within Γ_ν for some μ , there exists M such that

$$\varepsilon_n \le M/R^n$$

for all n .

Proof It is known that ψ is continuous on Γ ([9], p.86). On Γ and for any suitable branch of the complex logarithmic function we have

$$\ell n(\psi(z)) = \ell n(\frac{\psi(z)}{g(z)}) + \ell n(g(z)) \quad .$$

Taking the real part and recalling that $|\psi(z)| = 1$, $z \in \Gamma$, we obtain

$$\ell n|g(z)| = - \operatorname{Re}(A(z)) \quad \text{where} \quad A(z) = \ell n(\frac{\psi(z)}{g(z)}) \quad .$$

Hence

$$\varepsilon_n \leq \|\operatorname{Re}(-A(z)-q(z))\| \leq \|-A(z)-q(z)\|$$

for any polynomial q of degree n .

The results follow by applying the appropriate approximation theorems from [4]. (see pp.36, 75).

The estimation in (b) where applicable appears to be a good estimate of the rate of convergence in practice, and provides an aid to choosing g .

Theorems 1 and 2 provide the basic justification for the method and we now proceed to a more detailed analysis. Two results lie immediately to hand.

THEOREM 3

Let

$$G_n(z) = \ell n|g(z)| - \operatorname{Re}(p_n(z)) \quad .$$

Then

$$|G(z) - G_n(z)| \le \varepsilon_n \quad \text{for all} \quad z \in \bar{D} - \{0\} \quad .$$

Proof This is a simple consequence of the maximum principle, noting that $G-G_n$ is harmonic on D, the singularity at zero being removable.

THEOREM 4

For each $z \in \bar{D}$

$$\exp(-\varepsilon_n) \le \left| \frac{\phi_n(z)}{\psi(z)} \right| \le \exp(\varepsilon_n) \quad .$$

Proof $\left| \frac{\phi_n(z)}{\psi(z)} \right|$ achieves its maximum and minimum values on Γ: these values are $\exp(\varepsilon_n)$ and $\exp(-\varepsilon_n)$.

We note that $\exp(\pm\varepsilon_n) = 1 \pm \varepsilon_n + O(\varepsilon_n^2)$ so the image of Γ is within $O(\varepsilon_n^2)$ of the "best approximation" to the unit circle. For most values of ε_n under consideration the $O(\varepsilon_n^2)$ term is negligible. Thus ε_n is directly comparable with the quantity E_n given by Levin, et al. in [3] which measures the maximum deviation from the circle of the image of Γ given by their expansion.

To demonstrate convergence of the approximate mapping functions $\phi_n(z)$ to ψ is rather more difficult and we have been unable as yet to demonstrate uniform convergence on $\bar{D}$. However we have

THEOREM 5

The sequence (ϕ_n) converges to ψ on D, uniformly on compact subsets.

Proof Consider $\theta_n(z) = \phi_n(\psi^{-1}(z))$. These functions are analytic and uniformly bounded on the unit open disc S so they form a normal family. (see e.g. [7], chapter 15.) Thus each subsequence of the sequence (θ_n) has a further subsequence (θ_{n_j}) which convergence on S uniformly on compact subsets, to some limit function, say $\hat{\theta}(z)$. $\hat{\theta}$ must be the identity function on S , for by Schwarz Lemma (see e.g. [7], p.235) we see that $\hat{\theta}(z) = e^{i\alpha}z$ for some real α . But $\hat{\theta}'(0)$ is real and positive so $\hat{\theta}(z) = z$.

It follows that the whole sequence (θ_n) converges, uniformly on compact subsets, to the identity function on S and the result required is obtained by noting that $\psi(w) = z$.

It seems plausible that a sufficient condition for uniform convergence on D is that ψ/g be analytic and non-zero on an open set containing $\bar{D}$, but this has yet to be proved.

III. COMPUTED EXAMPLES

The approximations were computed by Lawson's algorithm (see e.g. [6]) using 100 points equally spaced with respect to the parameterisation of Γ chosen and with additional points where Γ was not smooth. This method was used as a program was available, but other methods given in [5] might well be more efficient. Where possible, symmetry was exploited to reduce the dimension of the problem: if Γ is symmetrical about the real axis the imaginary parts can be neglected and similarly symmetry about the imaginary axis means that only even powers of z are required.

Where only one value of ε_n for a particular example is given, this is the value for the largest successful n (subject to the remark in the previous sentence). In most cases the approximation program failed on the next n .

The following abbreviations are used when comparing with

other methods:

Symm : Symm's integral equation method[2].
BKM : The Bergman Kernel method with polynomial basis[3].
BKM/AB : The Bergman Kernel method with augmented basis[3].

The numbers given for these are the best E_n for the Bergman Kernel methods and the equivalent quantity for Symm's method.

The computations were carried out on the ICL 1904A computer at Brighton Polytechnic and it should be noted that this machine has a considerably shorter word length than the CDC 7600 used by Levin et al. in [3]. This precluded a close comparison of the two methods, as both are limited by the largest practicable value of n that could be used on the machine.

The diagrams were produced on an ICL 1934/6 graph plotter attached to the 1904A.

Example 1 The Unit Square

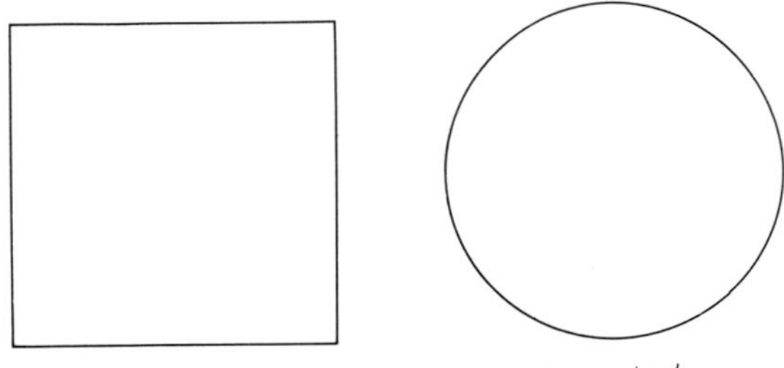

$g(z) = z$. $\varepsilon_{32} = 2.47\text{x}10^{-10}$.

(BKM: $1.4\text{x}10^{-8}$, BKM/AB: $3.4\text{x}10^{-11}$.)

With $g(z) = z/(z^4-16)$ (taking account of the nearest poles to Γ), $\varepsilon_{16} = 5.82\times10^{-9}$.

Example 2 Egg Shape

$$\Gamma = \{z = x+iy \mid x = \alpha(\theta) \cos(\theta) , \; y = \alpha(\theta) \sin\theta , \; 0 \le \theta < 2\pi\}$$

where $\alpha(\theta) = \frac{1}{2} + \cos^{16}(\frac{\theta}{2})$.

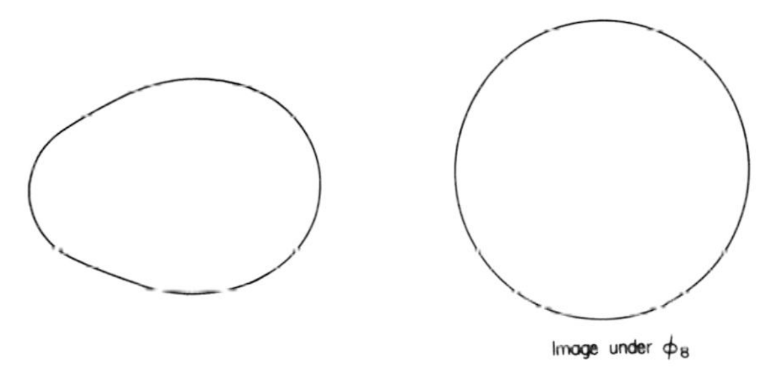

$$g(z) = z \; , \quad \varepsilon_8 = 5.31\times10^{-3} \; .$$

Example 3 "S"-Shape

$$\Gamma = \{z = x+iy \mid x = 2 \cos(\theta) , \; y = \sin(\theta) + 2 \cos^3(\theta) , 0 \le \theta < 2\pi\}$$

$$g(z) = z \; . \quad \varepsilon_6 = 4.19\times10^{-2} \; . \quad \varepsilon_8 = 2.96\times10^{-2} \; .$$

Note that ϕ_n is not 1 - 1 on Γ : this sometimes seems to occur on regions with long, narrow tongues. Neither Symm nor Levin et al. provide plots of their mappings, so it is not clear whether this problem occurs with their methods.

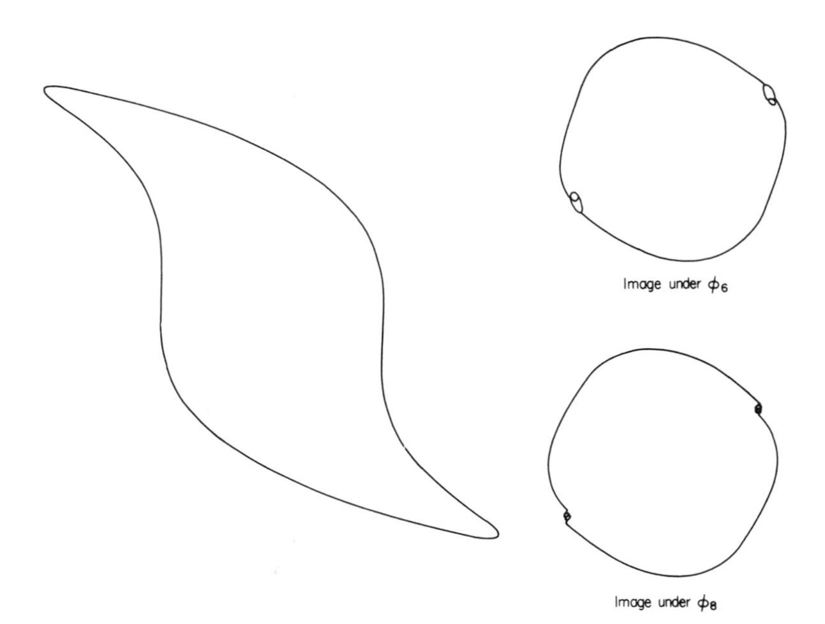

Example 4 Ellipse

$$\Gamma = \{z = x+iy \,|\, x = a\cos(\theta)\ ,\ \ y = \sin(\theta)\ ,\ \ 0 \le \theta < 2\pi\}\quad .$$

This example provides a convenient test of the effect of zeros and poles on the approximation, since ψ is known explicitly. (See [8], p.177.) On the other hand it is a somewhat artificial example as such detailed information is unlikely to be available in practical cases.

The poles and zeros of ψ are interlaced along the imaginary axis, those nearest to Γ being at $\pm ip$ and $\pm iq$ respectively, where

$$p = 2a/\sqrt{a^2-1} \quad \text{and} \quad q = 4a(a^2+1)/(a^2-1)^{3/2} \quad .$$

The results given below relate to the case $a = 5$.

(a) $g(z) = z$. $\varepsilon_{30} = 1.33\text{x}10^{-4}$ (BKM: $1.0\text{x}10^{-2}$)

(b) $g(z) = z/(z^2+p^2)$, $p = 2a/\sqrt{a^2-1}$.

$\varepsilon_{26} = 2.74\text{x}10^{-9}$ (BKM/AB: $3.3\text{x}10^{-10}$)

(c) $g(z) = z(z^2+q^2)/(z^2+p^2)$.

$p = 2a/\sqrt{a^2-1}$, $q = 4a(a^2+1)/(a^2-1)^{3/2}$.

$\varepsilon_{16} = 1.31\text{x}10^{-9}$.

Symm reports a typical error with $a = 2$ of $3\text{x}10^{-4}$.

Example 5 L-Shape

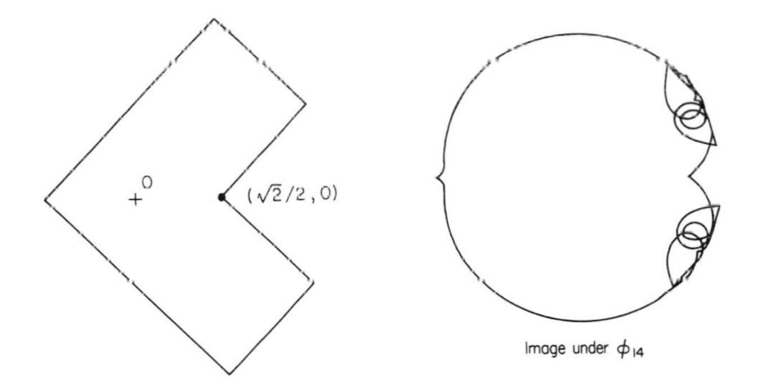

Image under ϕ_{14}

The first derivative of ψ is infinite at $(\sqrt{2}/2,0)$ and convergence is extremely slow with $g(z) = z$. In fact the values obtained were $\varepsilon_4 = 2.33\text{x}10^{-1}$, $\varepsilon_8 = 1.62\text{x}10^{-1}$,

$\varepsilon_{14} = 1.36\text{x}10^{-1}$, $(\varepsilon_{15} = 1.38\text{x}10^{-1})$, (BKM: $1.9\text{x}10^{-1}$, Symm: $4.1\text{x}10^{-2}$) .

The convergence of ε_n is apparently sublinear. (Compare Theorem 1 and the remark following on p.371 of [4].)

To remove the singularity in the first derivative, we take

$$g(z) = (z-z_0)^{2/3} - (-z_0)^{2/3} \quad \text{with} \quad z_0 = \sqrt{2}/2 \quad .$$

(Compare [3] p.7.) For convenience the cut was taken along the positive real axis to the right of z_0 , so $g'(0)$ is not real, thus introducing a rotation. However the symmetry of $\ell n|g(z)|$ is not affected. This choice of g gives $\varepsilon_6 = 3.98\text{x}10^{-2}$, $\varepsilon_{12} = 1.33\text{x}10^{-2}$, $\varepsilon_{16} = 1.19\text{x}10^{-2}$, $\varepsilon_{18} = 1.05\text{x}10^{-2}$. (The constraints of the system precluded a plot of ϕ_{18} .)

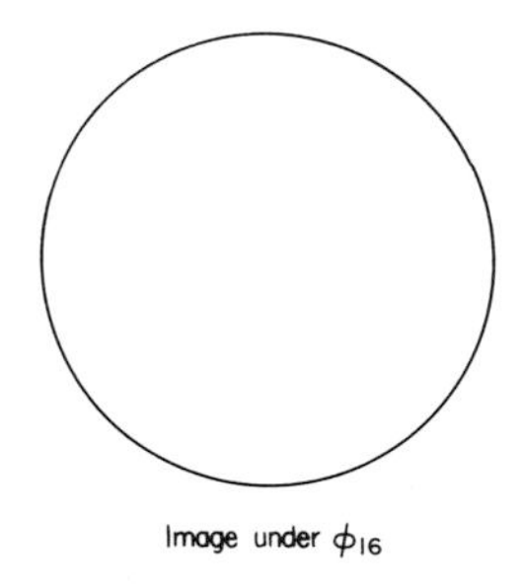

Image under ϕ_{16}

Levin et al. using BKM/AB eliminated singularities in higher derivatives and obtained a value of E_n of $2.2\text{x}10^{-5}$. To eliminate the discontinuity in the second derivative in the method presented here would appear to make the approximation problem non-linear as g would have to be allowed to vary: however this might be a fruitful line for further development.

Example 6 An Exterior Mapping Problem

It is possible to approach the problem of mapping the outside of Γ to the outside of the unit circle by approximating $\ell n|g(z)|$ by a polynomial in $\frac{1}{z}$. The diagram shows a typical result for the ellipse

$$\Gamma = \{z = x+iy \mid x = 2\cos(\theta) ,\ y = \sin(\theta) ,\ 0 \le \theta < 2\pi\} .$$

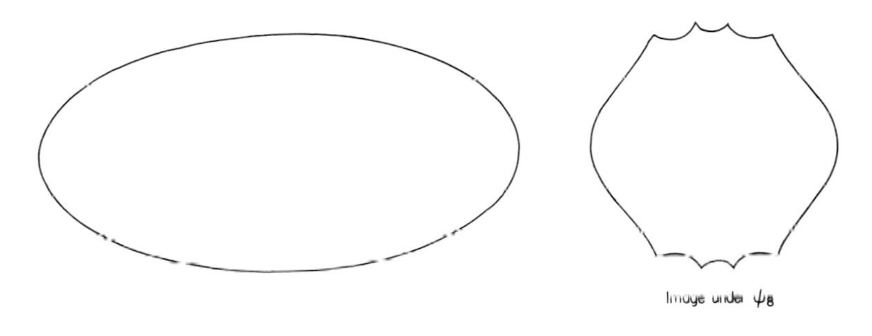

$$g(z) = z \quad . \quad \varepsilon_8 = 5.28\text{x}10^{-2} .$$

Note that the mapping has critical points on the boundary: this appears to be a typical feature of the application of the method to this type of problem.

IV. CONCLUSION

A method has been presented which appears to give comparable accuracy with existing methods and gives the approximate mapping in a particularly convenient form. Moreover the method is sufficiently flexible to allow considerable development. For instance g could be allowed to vary, or alternative basis functions could be used.

ACKNOWLEDGEMENT

The author would like to thank Dr J. Wood for several

helpful discussions and in particular for suggesting the proof of Theorem 1.

REFERENCES

1. Gram, C. (Ed.) (1962). "Selected Numerical Methods", Regnelentralen, Copenhagen.

2. Symm, G.T. (1974). *in* "Numerical Solution of Integral Equations (Ed. L.M. Delves and J. Walsh)" Oxford: Clarendon Press.

3. Levin, D., Papamichael, N. and Sideridis, A. (1977). "The Bergman Kernel Method for the Numerical Conformal Mapping of Simply Connected Domains", Brunel University, Dept. of Maths. TR/71.

4. Walsh, J.L. (1969). Interpolation and Approximation by Rational Functions in the Complex Domain, *Amer. Math. Soc. Coll. Pub.* XX 5th ed.

5. Andreasson, D.O. and Watson, G.A. (1976). "Linear Chebyshev Approximation without Chebyshev Sets", Dundee University Technical Report.

6. Rice, J.R. (1969). "The Approximation of Functions (vol. II)", Addison-Wesley.

7. Hille, E. (1962). "Analytic Function Theory (Vol.II)", Ginn.

8. Kober, H. (1957). "Dictionary of Conformal Representations", Dover, New York.

9. Carathédory, G. (1952). "Conformal Representation", Cambridge University Press, 2nd Ed.

A MIXED-NORM BIVARIATE APPROXIMATION PROBLEM WITH APPLICATIONS TO LEWANOWICZ OPERATORS

E.W. Cheney[†*], J.H. McCabe[††], G.M. Phillips[††]

[†]*Department of Mathematics*
University of Texas at Austin
Austin, Texas, U.S.A.

[††]*Mathematical Institute*
University of St. Andrews
St. Andrews, Scotland

1. INTRODUCTION

We consider here a class of approximation problems having the following form. A convex set F of bivariate functions $f(x,y)$ is prescribed, and it is desired to determine the quantity

$$\zeta = \inf_{f\in F} \max_{-1\leq x\leq 1} \int_{-1}^{1} |f(x,y)|\,d\mu(y) \quad .$$

Here μ is a prescribed measure. It is desired also to obtain a "minimum solution", that is, a particular element f_0 in F for which

$$\max_{-1\leq x\leq 1} \int_{-1}^{1} |f_0(x,y)|\,d\mu(y) = \zeta \quad .$$

It will be observed that this problem involves both a supremum norm and an L_1-norm.

One source of such approximation problems is in the study of integral transforms. It is sometimes necessary to approximate one integral transform by another one which is simpler.

---

*This author was supported by the Science Research Council of the U.K. and the U.S. Army Research Office.

If the first has the form

$$(Lf)(x) = \int_{-1}^{1} K(x,y)f(y)d\mu(y)$$

and if the second has the form

$$(L'f)(x) = \int_{-1}^{1} G(x,y)f(y)d\mu(y)$$

then a measure of the distance between L' and L is given by

$$||L'-L|| = \sup_{-1\le x\le 1} \int_{-1}^{1} |G(x,y) - K(x,y)|d\mu(y) \quad .$$

A specific application of this nature is considered in Section 3 below.

2. THE MAIN THEOREM

In this section we formulate and prove a theorem which characterizes the solution of the problem outlined in the previous section.

Let X be a compact Hausdorff space, and let (Y,Σ,μ) be a finite measure space. Let $\mathcal{F}$ be a prescribed convex set of functions from $X\times Y$ to $\mathbb{R}$. About $\mathcal{F}$ we make the following assumptions.

(1) *Equicontinuity* For each $f \in \mathcal{F}$, $\{f(\cdot,y): y \in Y\}$ is an equicontinuous family in $C(X)$. This means that

$$\lim_{\delta\to 0} \sup_{y\in Y} \sup_{|x-u|<\delta} |f(x,y) - f(u,y)| = 0 \quad .$$

(2) *Integrability* For each $f \in \mathcal{F}$ and for each $x \in X$,

$f(x,\cdot)$ belongs to the Lebesgue class $L_1(Y,\Sigma,\mu)$.

(3) *Smoothness* For each $f \in F$ and for each $x \in X$,

$$\mu\{y: f(x,y) = 0\} = 0 \quad .$$

Thus the zeros of $f(x,\cdot)$ form a set of measure 0 . In particular, $0 \notin F$. The term "smoothness" is used here in accordance with established meaning in normed linear spaces. Thus, in a normed linear space, an element f is a *smooth* element if there is a unique hyperplane of support at f to the ball of radius $\|f\|$.

For each $f \in F$ there is a "set of critical points" defined by

$$K(f) = \{x: \int |f(x,y)|dy = \sup_t \int |f(t,y)|dy\} \quad .$$

MAIN THEOREM *For an element* f_0 *of* F *, the following two conditions are equivalent:*

(A) $$\sup_x \int |f_0(x,y)|dy \le \sup_x \int |f(x,y)|dy \quad \textit{for all} \quad f \in F \quad .$$

(B) $$\inf_{x \in K(f_0)} \int \{f_0(x,y) - f(x,y)\} \operatorname{sgn} f_0(x,y)\,dy \le 0 \quad \textit{for all} \quad f \in F \;.$$

Proof The idea of the proof is to turn the given problem into a pure Tchebycheff approximation problem and then to apply the Kolmogoroff Criterium.

To this end, let V be the set of all functions v in $L_\infty(Y,\Sigma,\mu)$ for which $|v(y)| \le 1$ a.e. For each $f \in F$, we define $\tilde{f}$ on $X \times V$ by means of the equation

$$\tilde{f}(x,v) = \int v(y)f(x,y)\,dy \quad .$$

We give V the weak*-topology induced by $L_1(Y,\Sigma,\mu)$. Thus the convergence to v of a generalized sequence v_α is defined to mean

$$\lim_\alpha \int v_\alpha(y)g(y)\,dy = \int v(y)g(y)\,dy \quad \text{for all} \quad g \in L_1(Y,\Sigma,\mu) \quad .$$

It can now be proved that $\tilde{f}$ is continuous on $X \times Y$. Indeed if $(x_\alpha, v_\alpha) \to (x,v)$ then

$$\begin{aligned}
&|\tilde{f}(x_\alpha,v_\alpha) - \tilde{f}(x,v)| \\
&\quad = \left|\int v_\alpha(y)f(x_\alpha,y)\,dy - \int v(y)f(x,y)\,dy\right| \\
&\quad \le \left|\int v_\alpha(y)[f(x_\alpha,y)-f(x,y)]\,dy\right| + \left|\int v_\alpha(y)f(x,y)\,dy - \int v(y)f(x,y)\,dy\right| \\
&\quad \le \int |f(x_\alpha,y)-f(x,y)|\,dy + \left|\int v_\alpha(y)f(x,y)\,dy - \int v(y)f(x,y)\,dy\right| \quad .
\end{aligned}$$

The two terms on the last line converge to 0 by virtue of equicontinuity and by the meaning of weak*-convergence.

Next we observe that

$$\begin{aligned}
\sup_{x\in X} \int |f(x,y)|\,dy &= \sup_{x\in X}\sup_{v\in V} \int v(y)f(x,y)\,dy \\
&= \sup_{(x,v)\in X\times V} \tilde{f}(x,v) \quad .
\end{aligned}$$

The minimization of this last term is a Tchebycheff Approximation Problem. The compactness of V is a consequence of the Alaoglu Theorem[2], because $L_\infty(Y,\Sigma,\mu)$ is the dual of $L_1(Y,\Sigma,\mu)$ [[2], p.289].

According to the Kolmogorov Criterium (see the lemma below) the following conditions are equivalent:

(α) $\sup_{(x,v)\in X\times V} \tilde{f}_0(x,v) \leq \sup_{(x,v)} \tilde{f}(x,v)$ for all $f \in F$.

(β) $\inf_{(\xi,\eta)} (\tilde{f}_0-\tilde{f})(\xi,\eta) \leq 0$ for all $f \in F$.

The infimum in (β) is over all critical pairs (ξ,η) . The pair $(\xi,\eta) \in X\times V$ is *critical* if

$$||\tilde{f}_0|| = \tilde{f}_0(\xi,\eta) = \int \eta(y) f_0(\xi,y)\,dy \quad .$$

Since the zeros of $f_0(\xi,\cdot)$ form a set of measure zero, the sign function η must be almost everywhere equal to sgn $f_0(\xi,\cdot)$. Thus we see that conditions (α) and (A) are equivalent and that conditions (β) and (B) are equivalent. □

LEMMA (Kolmogoroff's Criterium) *Let* C *be a convex set of continuous real-valued functions defined on a compact Hausdorff space* T . *For each* $g \in G$, *let* $\Delta(g) = \max_{t\in T} g(t)$ *and* $K(g) = \{t \in T\colon g(t) = \Delta(g)\}$. *For any* $g_0 \in G$ *the following are equivalent:*

(A) $\Delta(g_0) \leq \Delta(g)$ *for all* $g \in G$.

(B) $\inf_{t\in K(g_0)} (g_0-g)(t) \leq 0$ *for all* $g \in G$.

The preceding lemma is a variation on a well-known result. See [3] p.15, for example. For the reader's convenience, we sketch the proof here. If (A) is false, select $g_1 \in G$ for which $\Delta(g_1) < \Delta(g_0)$. Then for $t \in K(g_0)$ we have $(g_0-g_1)(t) \geq \Delta(g_0) - \Delta(g_1)$. Conversely, if (B) is false, select $g_1 \in G$ and $\delta > 0$ so that $(g_0-g_1)(t) > \delta$ for all $t \in K(g_0)$. Put $S = \{t\colon (g_0-g_1)(t) > \delta\}$. Then $K(g_0) \subset S$. Since $T\setminus S$ is compact, the supremum, μ , of g_0 on $T\setminus S$ is attained and is less than $\Delta(g_0)$. For an appropriate $\lambda > 0$, we have $\lambda g_1(t) + (1-\lambda)g_0(t) = g_0(t) + \lambda(g_1-g_0)(t) \le \Delta(g_0) - \lambda\delta$ for $t \in S$. For $t \in T\backslash S$ we have $\lambda g_1(t) + (1-\lambda)g_0(t) = g_0(t) + \lambda(g_1-g_0)(t) \le \mu + \lambda\Delta(g_1-g_0)$. Thus $\Delta(\lambda g_1+(1-\lambda)g_0) < \Delta(g_0)$.

3. APPLICATION TO LEWANOWICZ OPERATORS

In [1], S. Lewanowicz has introduced a family of projection operators which map the space $C[-1,1]$ into the subspace Π_n of nth degree polynomials. His operators have the form

$$P = S_n + L \tag{3.1}$$

where S_n is the Fourier-Tchebycheff operator, and L is an operator which maps $C[-1,1]$ into Π_n and maps Π_n into 0 . The formulae for these operators are as follows

$$S_n = \sum_{k=0}^{n}{}' a_k \otimes T_k \tag{3.2}$$

$$a_k(f) = \frac{2}{\pi}\int_0^{\pi} f(\cos\theta)\cos k\theta \, d\theta \tag{3.3}$$

$$T_k(x) = \cos(k \arccos x) \tag{3.4}$$

$$L = \sum_{m \ge k > n} a_k \otimes p_k \quad . \tag{3.5}$$

Here p_k can be arbitrary elements of Π_n . The sum with the dash has its first term multiplied by $1/2$. The tensor notation means for example

$$(S_n f)(x) = \sum_{k=0}^{n}{}' a_k(f) T_k(x) \quad . \tag{3.6}$$

Lewanowicz proposed the problem of selecting the polynomials $p_{n+1}, p_{n+2}, \ldots, p_m$ in such a way that the norm of P

is as small as possible. He proved that if $p_{n+2} = p_{n+3} = \dots p_m = 0$ then $\|P\| \geq \|S_{n/2}\|$ when n is even. In this case, he showed that a good choice for p_{n+1} is $(2n)^{-1}U_{n-1}$, where U_{n-1} is the Tchebycheff polynomial defined by $U_{n-1}(\cos\theta) = \sin n\theta/\sin\theta$.

The minimization of $\|P\|$ by choosing appropriate polynomials $p_{n+1},\dots,p_m$ is a problem to which the theorem of Section 2 above is applicable. It is necessary first to note that

$$\|P\| = \sup_{\|f\|_\infty \leq 1} \|Pf\| \tag{3.7}$$

$$= \sup_{-1\leq x\leq 1} \int_{-1}^{1} \Big| {\sum_{k=0}^{n}}' T_k(x)T_k(y) + \sum_{k=n+1}^{m} T_k(y)p_k(x) \Big| \frac{dy}{\sqrt{1-y^2}}$$

Thus, in order to apply the theorem of Section 2, one takes $X = Y = [-1,1]$ and

$$f(x,y) = {\sum_{k=0}^{n}}' T_k(x)T_k(y) + \sum_{k=n+1}^{m} T_k(y)p_k(x) \tag{3.8}$$

$$\mu(a,b) = \arccos a - \arccos b \quad . \tag{3.9}$$

The family F consists of all the f that are obtained in Equation (3.8) when p_k range independently over Π_n. The following theorem results from applying the main theorem.

THEOREM *Let* f^* *be the result of making the specific choice* $p^*_{n+1},\dots,p^*_m$ *in Equation (3.8). In order that this be an optimal choice for minimizing* $\|P\|$ *it is necessary and sufficient that*

$$\inf_{x\in K} \sum_{k=n+1}^{m} p_k(x)g_k(x) \leq 0$$

for all choices of $p_{n+1},\ldots,p_m$ *in* Π_n . *Here* K *is the set of points where*

$$\int_{-1}^{1} |f^*(x,y)|(1-y^2)^{-1/2}\,dy$$

attains its maximum, and

$$g_k(x) = \int_{-1}^{1} T_k(y)\,\text{sgn}\,f^*(x,y)(1-y^2)^{-1/2}\,dy \quad .$$

EXAMPLE As an illustration of the main theorem, consider the problem of minimizing the expression

$$\max_{-1\le x\le 1} \int_{-1}^{1} \Big| \sum_{k=0}^{n}{}' T_k(x)T_k(y) - T_{n+1}(y)(1-x^2)p(x) \Big| \frac{dy}{\sqrt{1-y^2}}$$

by choosing p in Π_{n-2} . One solution is obtained by taking $p(x) = 0$. Indeed, the value of the integral at $x = \pm 1$ is independent of p , and is a maximum value when $p = 0$. The condition of the theorem is met because there are only two critical points, $x = \pm 1$, and at each critical point x ,

$$\int T_{n+1}(y)\,\text{sgn}\Big[\sum_{k=0}^{n}{}' T_k(x)F_k(y)\Big] \frac{dy}{\sqrt{1-y^2}} = 0 \quad .$$

We have carried out numerically the optimization of Lewanowicz's operator with a single parameter. Namely, we considered

$$Q_n = S_n + rn^{-1}a_{n+1} \otimes U_{n-1}$$

with r chosen to minimize the norm of Q_n . The theorem of Section 2 applies to this case, and the result is as follows:

THEOREM *In order that* $\|Q_n\|$ *be minimized by a particular* *choice* r^* *of the real parameter* r , *it is necessary and sufficient that the function*

$$\phi(x) := U_{n-1}(x) \int_{-1}^{1} T_{n+1}(y) \, \mathrm{sgn}\{ \sum_{k=0}^{n}{}' T_k(x) T_k(y) + r^* n^{-1} T_{n+1}(y) U_{n-1}(x) \} \frac{dy}{\sqrt{1-y^2}}$$

satisfy $\inf_{x \in K} \phi(x) \le 0$ and $\sup_{x \in K} \phi(x) \ge 0$.

The table which follows shows, for a variety of values of n , the optimal values of r and the corresponding values of $\|Q_n\|$.

n	r	$\|Q_n\|$	n	r	$\|Q_n\|$
1	0.5	1.273	9	0.442	1.930
2	0.5	1.436	10	0.443	1.968
3	0.5	1.552	11	0.443	2.003
4	0.484	1.643	12	0.448	2.035
5	0.462	1.718	13	0.444	2.066
6	0.451	1.782	14	0.446	2.094
7	0.448	1.837	15	0.458	2.117
8	0.449	1.885	20	0.477	2.224

REFERENCES

1. Lewanowicz, S. (1976). "A Projection Connected with the Fourier-Chebyshev Operator", Preprint, Institute for Informatics, University of Wrocław, Poland.
2. Dunford, N. and Schwartz, J.T. (1958). "Linear Operators, Part I", Interscience Publishers, New York.
3. Meinardus, G. (1967). "Approximation of Functions", Springer.

MULTIVARIATE APPROXIMATION PROBLEMS IN COMPUTATIONAL GEOMETRY[*]

A.R. Forrest

School of Computing Studies
University of East Anglia
Norwich

1. INTRODUCTION

Computational Geometry[12,27] is concerned with the handling, by computer, of shape information. There are many difficulties in the processing of geometric data, not the least of which is the problem of mathematical representation of two and three dimensional objects. The paper sets out to highlight some of these difficulties, with particular reference to multivariate approximation.

Historically, the greater part of the effort in computer-aided geometric design[2,8,11] has been concerned with the representation of the so-called sculptured surfaces as exemplified by ship hulls, car bodies and aircraft exterior skins. Here the problem is to develop, preferably with the aid of a designer who is not a mathematician or computer scientist, a mathematical representation of a basically smooth surface. The representation should be compact, simple to handle, and simple to analyse.

[†] This research partially supported by SRC Grant B/RG/95834, Investigations in Computational Geometry.

An obvious source of potentially useful techniques is approximation theory; but it is not always apparent to the theorists that methods appropriate to the approximation of *functions* are not necessarily applicable to the approximation of *shapes*. Most shapes are not functions in the strict sense of the word - rather they are *vector-valued functions* if they are functions at all. The use of vector-valued functions to approximate shape brings the advantages of axis independence, the ability to handle tangents in any direction, and the ability to model re-entrant shapes without recourse to tedious changes of axes. Unfortunately there is little in the way of a theory for the approximation of vector-valued functions and many of the methods currently used by industry are ad hoc. However, the practical achievements are sometimes impressive, and have on occasion led to subsequent theoretical insight into the reasons for success[13,16].

In this paper, we discuss first the role of *variation diminishing* approximation methods[12,32] in the design and modelling of three dimensional objects. We then consider the need for *structure* (in the sense of a data structure and a geometric structure) in modelling three dimensional objects, and finally we explore how the use of variation diminishing techniques and structures can lead to efficient and reliable algorithms for the analysis of models of three-dimensional objects.

2. VARIATION DIMINISHING METHODS

Computer-aided geometric design should involve the designer in a natural and convergent series of decisions which lead to the construction of a satisfactory model of a geometric object. The correctness of the model can best be determined by *the designer himself*. No automatic procedure can guarantee to produce satisfactory results for all input data; "best approximation" depends on the designer, the application, the time of day, etc.; hence the need for positive feedback from the user to the modelling system.

The designer wishes to create his model quickly and predictably with the minimum of data. His interface with the model should be natural to him - for example, a ship hull designer should not need to know that the B-spline basis[6,9] has been used in constructing his hull. It is, however, essential that the handles by means of which he manipulates his model should have predictable and sensible effects.

Perhaps the most natural and most obvious handle for the design of surfaces is a point (strictly a point vector) through which the surface is constrained to pass. By controlling a point on the surface, the designer can affect the shape of the surface *directly*. Three issues must be decided by the system designer: how many points should the user be allowed to define his shape; how should these points be arranged on the surface (regularly or randomly); and how should the movement of a point affect the shape of the surface?

As a sound rule of thumb, the fewer points a designer has to specify, the smoother will be the resulting surface. It is a common failing to succumb to the temptation to provide the user with many control points in an attempt to capture the precise nuances of shape. This can lead to the necessity of performing complex smoothing operations in order to reduce inflexions unwittingly produced by the designer or his data.

Constraining control points to be regularly distributed on, say, a topologically rectangular grid, rather than randomly distributed, leads to compact representation in the computer and to important benefits which accrue when algorithms for geometric analysis are to be used. Of course, there are situations where regularly structured data are not available, and we shall return to this problem when we consider structure. By and large, if a structure can be imposed, then a regular rectangular distribution of points is to be preferred. There is an elegant theory for surfaces so defined[15] in contrast to the current lack of a coherent theory for, say, regular triangular point distributions (see [1] for a survey of the current state-of-the-art). Whereas the user can select from several families of rectangular patches, triangular patches do not fall into obvious categories.

The issue which most directly affects the designer is how the movement of a control point affects the shape of a surface. Change may be local or global, and ideally the user should be able to specify in a simple manner the region over which a change is to take place. Change may also be variation increasing or variation diminishing. Figure 1 illustrates such changes for curves; the corresponding changes for surfaces are obvious. Most designers would find the variation increasing methods unacceptable, since points of inflexion (ripples) are unexpectedly and unpredictably introduced. A designer, for example, would not wish a surface representation to include inflexions which were not obviously inherent in his data. Interpolation, in general, is a variation increasing procedure - the Runge effect in polynomial interpolation is well known; but the same effect is present in *all* interpolation with the exception of piecewise-linear interpolation, and the interpolation of some locally monotonic and/or convex data[20], as in Figure 1b. Piecewise linear interpolation reproduces the inherent variation of the data; variation increasing methods have a *lower* bound on variation which is the inherent variation, whereas variation diminishing methods have the inherent variation as an *upper* bound. However, variation diminishing

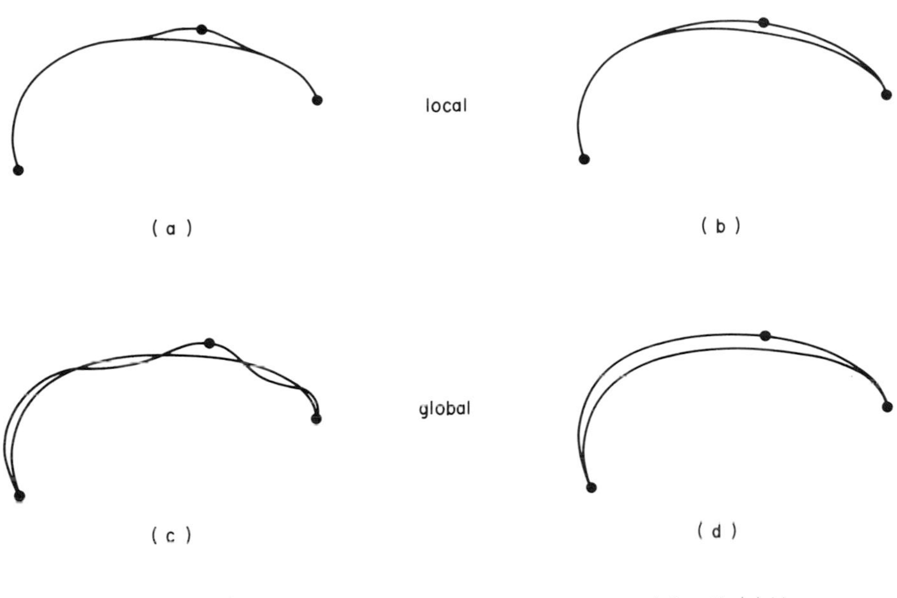

Figure 1

methods do not preserve the interpolation property at all points, except for particular configurations of data. Used locally, or semi locally, the effects of change can be confined.

Perhaps the most successful surface design system, in terms of user acceptability, has been the UNISURF system developed by Bézier at Renault[3]. Here the designers do not control points on a surface directly, but manipulate the vertices of a polygonal mesh[4,13] (Figure 2). The resulting surfaces are bivariate vector-valued Bernstein polynomial approximations to notional surfaces defined only at the vertices of the mesh. The resulting surfaces are thus smoothed versions of the defining mesh. By adopting such a technique, it is possible to guarantee that surfaces will be basically smooth and that "bumps" and

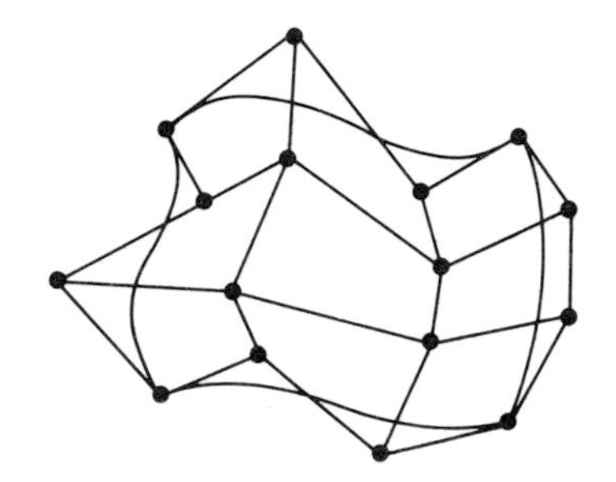

Figure 2 Polygonal mesh and corresponding surface

"hollows" will only occur where the designer has made a conscious decision to place them. More recently, the B-spline basis[6,9] (the obvious spline counterpart to the Bernstein polynomial basis) has been used to create systems which are globally variation diminishing but permit local changes[24]. (The Bézier methods are variation diminishing on a local - single polynomial - scale, but do not *directly* permit global control.) Several surface modelling systems currently under development in industry use the variation diminishing B-spline approach. For fuller details of the *geometric* properties of the B-spline basis, see Riesenfeld[24].

3. STRUCTURE

A major problem in defining surfaces in terms of random point sets is that interpolation cannot always be guaranteed[10]. However, if some ordering can be imposed on the data, then surface fitting becomes easier. If a triangulation can be found for the data, then there are several methods for surface fitting which are appropriate[21].

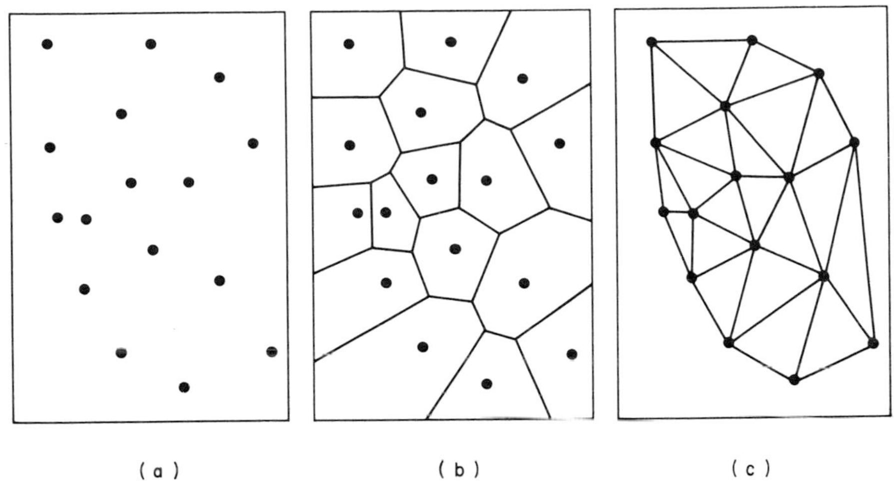

Figure 3 Triangulation of random data

Until recently, algorithms for triangulation did not necessarily terminate satisfactorily for all data, and even if a triangulation could be found, it was often by no means ideal. Sibson[30] has recently proved that the Delaunay triangulation (the dual graph of the Voronoi or, Dirichlet, tessellation of a random point set) is an optimal triangulation in the sense of Lawson[19]. The Voronoi tessellation is easily defined: each data point is surrounded by a polygon with the property that the interior points of the polygon are nearer the surrounded points than any other points in the data set. Mathematically, the region V_i surrounding the point P_i is the intersection of all the half planes $h(P_i, P_j)$ containing P_i and defined by the perpendicular bisector of $P_i - P_j$:

$$V_i = \bigcap_{i \neq j} h(P_i, P_j)$$

The mathematical definition does not lead directly to an

obvious algorithm for construction; but three recent papers[28,29,31] have provided practical algorithms. Figure 3 shows a random point set, the corresponding Voronoi polygons, and the dual triangulation. Imposition of a structure, in this case an irregular triangulation, enables surface fitting to be performed. It should be realised that obtaining such a structuring is by no means simple[1], since there is no natural ordering for points in three dimensions. Once the surface has been defined, the structure provides a convenient access mechanism for further analysis, e.g. the extraction of areas, plane sections, etc.

The need for structure where surfaces are defined by regularly distributed data is not readily apparent. In the past, the computational geometers have concentrated on surface *patches* or regular arrays of surface patches[2,8,11], ignoring the fact that in many cases, the surfaces are the boundaries of *volumes*. For reasons mentioned earlier, most surface systems use rectangular patches. However, fitting an array of rectangular patches is not necessarily the best approach. In two earlier surface fitting exercises, the author found that it was necessary to use degenerate patches (triangular patches formed by reducing one side of a rectangular patch to a point) in order to achieve a good *and natural* fit to the surface. Moreover in each case - a telephone handset and a shoe last - the triangular patch was *embedded* in an otherwise rectangular array. In a subsequent exercise concerning an aircraft windscreen[18], a five-sided region embedded in an array of four-sided patches seemed to be the most natural solution. The author is not alone in finding that non-regular arrays of patches are essential - Bézier[5] shows the use of 3 and 5-sided patches in car body design, and Sabin[25] has used both 3- and 5-sided regions for aircraft. Rice, in his paper at this symposium[23], gives an example from the analysis of a canine joint where a triangular element is embedded in an array of rectangular elements.

Attempts to dodge the issue by using only rectangular arrays of rectangular patches do not always lead to satisfactory results - in many cases it is not possible to maintain smoothness in the anomalous regions. The use of degenerate rectangular patches is unsatisfactory - again, a lack of smoothness is apparent, and there is always the question of which side to make degenerate. How then do we recognise the anomalous regions where special patches are needed? It is not merely a matter of topology, although the fact that a sphere cannot be covered by four-sided regions is significant. The author[14] believes that one answer is in the use of an underlying structure which indicates where anomalous regions are to be found. In Figure 4 we demonstrate how the rounding-off of the edges of simple polyhedral assemblies can lead to n-sided regions, and we show where equivalent configurations can arise in practice.

The anomalous regions occur when we round off the vertices of polyhedra. Hence, if we are rounding off a box, a triangular region will occur at the vertices where three edges intersect. One line of attack might be to rough out a design in terms of solid blocks and then smooth the blocks. Recall that one interpretation of Bézier's use of variation diminishing approximation is that he is rounding off or smoothing a polygonal mesh. If we position such polygonal meshes on the faces and edges of our blocks, then Bernstein or B-spline approximation to these meshes will produce the desired effect, *except at the vertices*. What we now need is a variation diminishing approximation to n-sided regions which is compatible with the conventional four-sided patches. Sabin[26] has recently developed a recursive algorithm which appears to have the required property;

but the method is *algorithmic* and an analytic representation of the n-sided region is not immediately available.

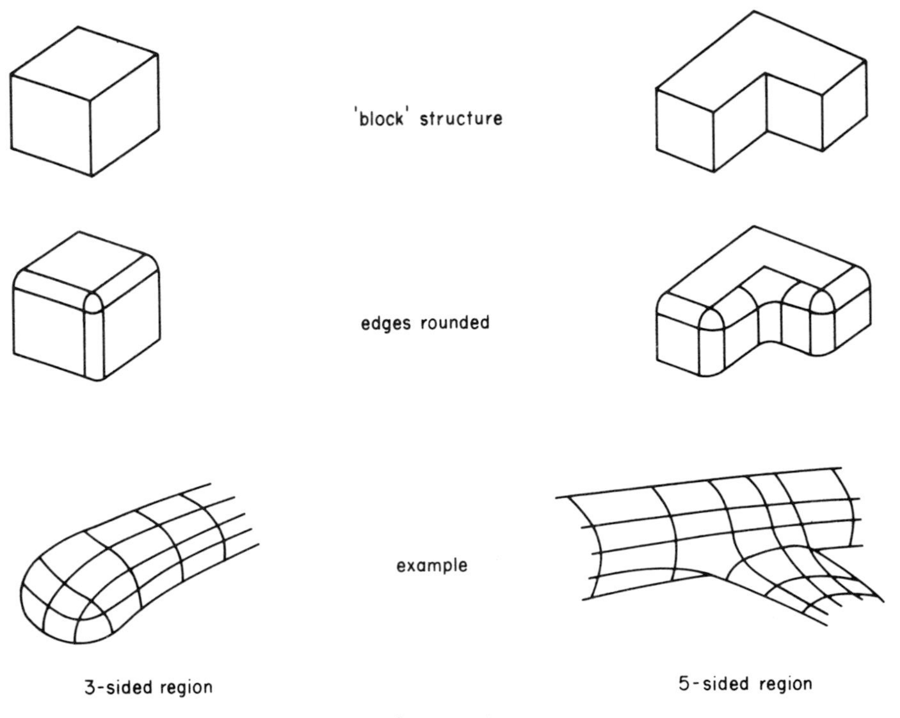

Figure 4

Current research at the University of East Anglia includes the development of a surface design system which is based on the BUILD system[7] for designing volumetric objects. The user will be expected to construct a block approximation to his desired object, and then to smooth the faces, edges and vertices of this *block structure* using variation diminishing techniques. Figure 5 shows a typical BUILD object and the suggested effect of an initial smoothing process.

Where a surface is to be fitted to predefined data, rather than constructed ab initio, the user may be requested to provide a block approximation to his data. If the data are random

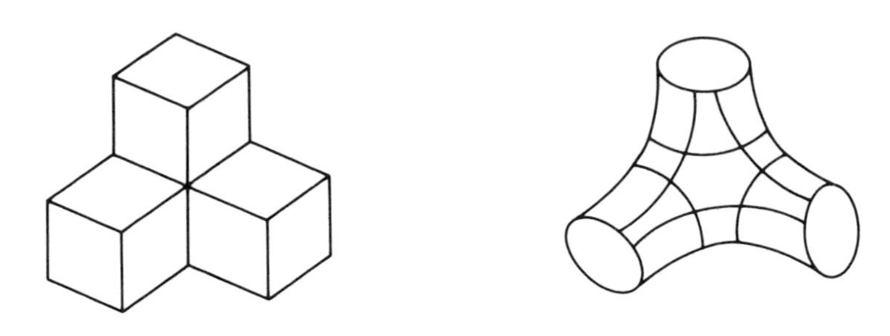

BUILD object Initial smoothing

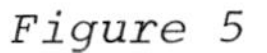

Figure 5

and sufficiently numerous, then an automatic procedure akin to edge extraction in visual scene analysis could be used. In a smoothed object, an "edge" is indicated by the two principal curvatures being markedly different in magnitude, with one (across the "edge") being large relative to values in surrounding regions. A localised differencing operator might be used to detect such features, and the "edges" and "vertices" so found would then be incorporated in a polyhedral approximation to the given data.

4. GEOMETRIC ALGORITHMS

Much emphasis has been placed on the mathematical *representation* of three-dimensional objects, and the question of how to *analyse* or interrogate such objects has received scant attention. Sabin is a notable exception[25]. The kinds of analysis we might wish to perform include: the determination of areas, enclosed volumes, whether two objects intersect, what is the shortest distance between two objects, etc. Many of the existing algorithms for the analysis of three dimensional shapes are unsatisfactory. Either they cannot guarantee to give correct results for all data, or they rapidly run out of time and space for objects of realistic and/or practical complexity. The combinatorial explosion is a serious problem. Until recent work at Yale University[28], the *inherent complexity* of various geometric operations was unknown. Some commonly used algorithms, for example, are $O(n^2)$ whereas $O(n \log n)$ algorithms are possible. We still do not know the complexity of many of the operations we might wish to perform. However, some operations are known to be "hard" and any heuristics we can devise will obviously be useful. The question of *completeness*, of being able to guarantee results, is important. For example, several widely used methods for determining plane sections of assemblies of surface patches will only detect intersections involving patch edges, missing top-of-the-mountain intersections. Moreover, if the search for edge intersections is conducted by a crude grid search, not all intersections will be detected. Clearly, if we wish to determine whether two objects will fit together, we cannot afford any uncertainty.

We propose to use variation diminishing methods, and block structure, to reduce or overcome these difficulties. By looking first at the block structure we can devise heuristics which will reduce the number of comparisons to be made in, say, detecting object-object interference. By exploiting the fact that variation diminishing approximations to polygons and polygonal meshes must lie within the convex hulls[22,24] of the polygons or polygonal meshes, we can eliminate many comparisons which might otherwise be made, and hence reduce the combinatorial problem.

A further benefit arises if our objects are variation diminishing approximations to block structures. Consider the problem of whether a given line intersects a given curve segment. If we know the polygon of which our curve is a variation diminishing approximation, then the intersection of the line with the polygon hull gives us an *upper bound* on the number of possible intersections. The tests for intersection are simple since we look in this case only at line-hull intersections. We proceed to line-curve intersections only if such intersections are indicated. Moreover, if we parametrise the polygon over the same parametric range used for the curve then the parameter values where line-polygon intersections occur are good approximate values for searching for line-curve intersections, Figure 6.

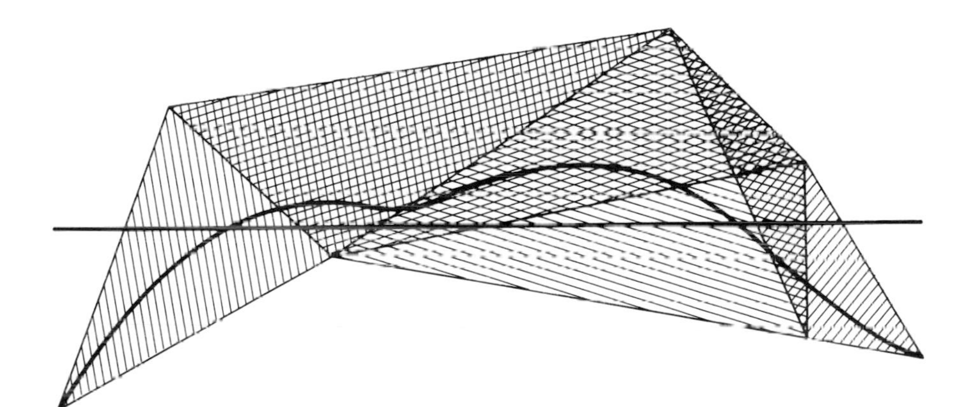

Figure 6 Line-curve intersections (cubic B-spline)

These ideas naturally generalise to surfaces - bivariate functions - in three dimensions; the problems become more difficult but there are considerable gains to be made over conventional methods, e.g. for determining surface-surface intersections.

We use variation diminishing approximation in a novel manner - in reverse: we derive objects whose variation diminishing approximations *are* the shapes we wish to model. The derived shapes will in general be simpler, and more angular, than the originals, but we can use the simplified shapes to perform both a pruning of the geometric objects to be analysed and to provide *approximations* which are appropriate starting points for more refined analysis.

5. CONCLUSIONS

We have discussed briefly some of the problems which have arisen in attempting to model three dimensional objects. These problems are not academic - they are of real practical importance - but nevertheless they pose intriguing theoretical problems. Approximation theory, albeit in a modified form, has provided some of the answers. In particular, variation diminishing techniques are seen to be of central importance, both for representation and for analysis. Much work remains to be carried out: in particular, variation diminishing approximations to n-sided regions, compatible with polynomial surfaces of arbitrary degree, need to be developed.

ACKNOWLEDGEMENTS

This research was partially supported by the Science Research Council Grant: Investigations in Computational Geometry, B/RG/958354. Figure 3 was generated by a development, by A.P. McRae, of the algorithm due to Sibson and Green[31]. The BUILD system[7] was developed by I.C. Braid, University of Cambridge, and made available to the Computational Geometry Project at the University of East Anglia.

REFERENCES

1. Barnhill, R.E. Representation and Approximation of Surfaces. Mathematical Software Symposium, University of Wisconsin, Madison, April 1977. To be published by Academic Press.

2. Barnhill, R.E. and Riesenfeld, R.D. (Eds.) (1975). Computer-Aided Geometric Design. Proceedings of a Conference, University of Utah, March 1974. Academic Press.

3. Bézier, P.E. (1972). Numerical Control - Mathematics and Applications. Wiley.

4. Bézier, P.E. (1977). Essai de Définition Numérique des Courbes et des Surfaces Experimentales. Thèse de Doctorat d'Etat, l'Université Pierre et Marie Curie.

5. Bézier, P.E. Mathematical and Practical Possibilities of UNISURF. In Barnhill and Riesenfeld, op. cit.

6. de Boor, C. (1972). On Calculating with B-Splines, *J. Approx. Theory* 6, 50-62.

7. Braid, I.C. (1975). The Synthesis of Solids Bounded by Many Faces, *Comm. ACM* 18, No.4.

8. Coons, S.A. (1967). Surfaces for Computer Aided Design of Space Forms, *M.I.T. Project MAC*, MAC-TR-41.

9. Cox, M.G. (1972). The Numerical Evaluation of B-Splines, *J. Inst. Maths. Applic.* 10, 134-149.

10. Davis, P.J. (1963). Interpolation and Approximation. Ginn-Blaisdell.

11. Forrest, A.R. (1968). Curves and Surfaces for Computer-Aided Design. Ph.D. Thesis, University of Cambridge.

12. Forrest, A.R. (1971). Computational Geometry, *Proc. Roy. Soc. Lond. A* 321.

13. Forrest, A.R. (1972). Interactive Interpolation and Approximation by Bezier Polynomials, *Computer Journal* 15, No.1.

14. Forrest, A.R. Computational Geometry - Achievements and Problems. In Barnhill and Riesenfeld, op. cit.

15. Gordon, W.J. (1969). Distributive Lattices and the Approximation of Multivariate Functions, *in* "Approximation with Special Emphasis on Spline Approximation", (Ed. I.J. Schoenberg) Academic Press.

16. Gordon, W.J. and Riesenfeld, R.F. (1974). Bernstein-Bezier Methods for the Computer-Aided Design of Free-Form Curves and Surfaces, *J. ACM* 21, No.2.

17. Lane, J.M. and Riesenfeld, R.F. (1977). The Application of Total Positivity to Computer-Aided Curve and Surface Design. Department of Computer Science, University of Utah, submitted for publication.

18. Lang, C.A. (1968). An Evening Designing with a Computer, *New Scientist* 43, No.660.

19. Lawson, C.L. Generation of a Triangular Grid with Applications to Contour Plotting. California Institute of Technology, Jet Propulsion Laboratory, Tech. Memo. 299.

20. McAllister, D.F., Passow, E. and Roulier, J.A. (1977). Algorithms for Computing Shape Preserving Spline Interpolations to Data, *Maths. of Computation* 30, No.139.

21. Powell, M.D. and Sabin, M.A. (1976). Piecewise Quadratic Approximations on Triangles. Dept. of Applied Mathematics and Theoretical Physics, Cambridge University.

22. Preparata, F.P. and Hong, S.J. (1977). Convex Hulls of Finite Sets of Points in Two and Three Dimensions. *Comm. ACM* 20, No.2.

23. Rice, J.R. Multivariate Piecewise Polynomial Approximation. These proceedings.

24. Riesenfeld, R.F. (1972). Applications of B-Spline Approximation to Geometric Problems of Computer-Aided Design. Ph.D. Thesis, Syracuse University. Published as University of Utah, *Computer Science*, UTEC-CSc-73-126, March 1973.

25. Sabin, M.A. (1977). The Use of Piecewise Forms for the Numerical Representation of Shape. Computer and Automation Institute, Hungarian Academy of Sciences, Report 60/1977. ISBN 963 311 035 1.

26. Sabin, M.A. (1977). Private Communication.

27. Shamos, M.I. (1974). Introduction to Computational Geometry. Yale University, Dept. of Computer Science.

28. Shamos, M.I. (1975). Geometric Complexity. Proc. 7th ACM SIGACT Conference, Albuquerque, N.M.

29. Shamos, M.I. and Hoey, D. (1975). Closest Point Problems. Proc. 16th IEEE FOCS Conference, Berkeley, Cal.

30. Sibson, R. Locally Equiangular Triangulations. To appear, *Computer Journal*.

31. Sibson, R. and Green, P.J. Computing Dirichlet Tessellations in the Plane. To appear, *Computer Journal*.

32. Stancu, D.D. Approximation of Bivariate Functions by Means of Some Bernstein-type Operators. These proceedings.

ONE-SIDED APPROXIMATION AND INTEGRATION RULES

H. Strauss

Institut für Angewandte Mathematik der
Universität Erlangen-Nürnberg
Erlangen, Germany

This paper deals with the relationship between the theory of one-sided approximation and minimal point integration rules.

Let an interval $B = [a,b]$ and $C^{(1)}[a,b]$ - the space of realvalued functions on $[a,b]$ with a continuous derivative - be given. Suppose $G \subset C^{(1)}[a,b]$ is a subspace of dimension r. For every function f we define the set of functions

$$G(f) = \{g \in G \mid g(x) \le f(x) \ , \ x \in [a,b]\} \quad .$$

The approximation problem studied here is:
Let $f \in C^{(1)}[a,b]$ be given. Determine the set of best approximations

$$P_{G(f)} = \{g \in G(f) \mid \int_a^b (f-g)(x)dx \le \int_a^b (f-h)(x)dx \quad \text{for} \quad h \in G(f)\} \ .$$

This is a problem of one-sided approximation. Results for one-sided approximation with Tchebycheff-systems were obtained by R. Bojanic and R. DeVore. First we extend these results.

We define the following sets of points which are very important:

Let $U = (u_i)_{i=1}^r \subset [a,b]$ be given. Then every function $g \in G$ satisfying $g(u_i) = 0$, $i = 1,\ldots,r$, shall vanish identically on the whole interval $[a,b]$. (1)

Further we denote by $Z(f) = \{x \in B \mid f(x) = 0\}$ the zeros of a function f.

THEOREM 1 Let $f \in C^{(1)}[a,b]$ be given. Suppose $g_o \in P_{G(f)}$ and there exists a set U such that $Z(f-h_o) \subset U$ where U satisfies condition (1). Then there exists an integration rule

$$Q(f) = \sum_{i=1}^{p} A_i f(u_i) \;, \; A_i > 0 \;, \; Z(f-h_o) = (u_i)_{i=1}^p$$

and Q is precise for G.

The theorem can be proved with the same arguments as Theorem 3.

Example 1 Let S be a set of splines with fixed knots; that is, every spline has the form

$$s(x) = \sum_{i=0}^{n-1} a_i x^i + \sum_{i=1}^{k} b_i (x-x_i)_+^{n-1} \;, \; a < x_1 < \ldots < x_k < b \;.$$

Let be

$$F = \{f \in C^{(n)}[a,b] \mid \zeta(-1)^i f^{(n)}(x) > 0 \;, \; x \in [x_{i-1},x_i] \;, \; \zeta = \pm 1 \;,$$
$$i = 1,\ldots,k+1 \;, \; x_o = a \;, \; x_{k+1} = b\} \;.$$

It can be proved that the function $f-s$, $s \in S$, has at most $n+k$ zeros. Let be $s_o \in P_{S(f)}$, $n+k$ even. It follows that the function $f-s_o$ has at most $p = (n+k)/2$ distinct zeros $(u_i)_{i=1}^p$. Now we can prove that there exists a set U such that $(u_i)_{i=1}^{p} \subset U$ where U satisfies condition (1). We conclude from Theorem 1 that there exists an integration rule Q

$$Q(f) = \sum_{i=1}^{p} A_i f(u_i) \quad , \quad A_i > 0$$

which is precise for S.
(See also [3,4].)

The integration rule of the example is called a minimal point rule since the number of free parameters of Q is equal to the dimension of S where Q is precise for all functions of S.

Now we extend our results to the case of two dimensions. Let B be a closed, bounded set in the plane with positive area and let $C^{(1)}(B)$ be the set of functions with a continuous derivative on B. Suppose $H \subset C^{(1)}(B)$ is a subspace of dimension r. For every function $f \in C^{(1)}(B)$ we define

$$H(f) = \{h \in H \mid h(x,y) \le f(x,y) \quad \text{for all} \quad (x,y) \in B\} \quad .$$

The set of best approximations is

$$P_{H(f)} = \{h \in H \mid \iint_B (f-h)(x,y)\,dxdy \le \iint_B (f-g)(x,y)\,dxdy \quad \text{for all} \quad g \in H(f)\} \quad .$$

We define again:

Let $U = (P_i)_{i=1}^{r} \subset B$ be given. Then every function $h \in H$ satisfying $h(P_i) = 0$, $i = 1,\dots,r$, shall vanish identically on the set B. (2)

THEOREM 2 Let $h_o \in P_{H(f)}$ be given. Suppose there exists a set U such that $Z(f-h_o) \subset U$ where U satisfies (2). Then there exists an integration rule

$$Q(f) = \sum_{i=1}^{p} A_i f(P_i) \ , \quad A_i > 0 \ , \quad (P_i)_{i=1}^{p} = Z(f-h_o)$$

which is precise for H .

Theorem 2 can also be proved in the same way as Theorem 3.

It is usually difficult to decide in this case whether $Z(f-h_o)$ is subset of a set satisfying (2) or not. For this reason we consider another approximation problem with half-discrete constraints.

Let R be a rectangle $R = \{(x,y) | a \le x \le b \ , c \le y \le d\}$ satisfying $B \subset R$. We may subdivide the rectangle by means of a grid. Let be $(z_i)_{i=0}^{t} \subset [c,d]$ satisfying

$$c \le z_o < z_1 < \ldots < z_t \le d \ , \quad t \le n \quad .$$

We define the grid

$$T = \bigcup_{i=0}^{t} T_i \ , \quad T_i = \{(x,y) \in B | y = z_i\}$$

and for every function f

$$H_T(f) = \{h \in H | h(x,y) \le f(x,y) \quad \text{for all} \quad (x,y) \in T\} \quad .$$

We study the following problem: Determine a function $h_o \in H_T(f)$ satisfying

$$\iint_B h(x,y)\,dxdy \le \iint_B h_o(x,y)\,dxdy \quad \text{for all} \quad h \in H_T(f) \ . \tag{3}$$

The existence of a function h_o depends also on the choice of the grid T .

THEOREM 3 Suppose there exists a solution h_o of problem (3) and a set U such that $Z(f-h_o) \cap T \subset U$ where U satisfies (2). Then there exists an integration rule

$$Q(f) = \sum_{i=1}^{p} A_i f(P_i) , \quad A_i > 0 , \quad (P_i)_{i=1}^{p} = Z(f-h_o) \cap T$$

which is precise for H .

Proof Let $(x_i,y_i)_{i=1}^{p}$ be the zeros $Z(f-h_o) \cap T$ and $(x_i,y_i)_{i=p+1}^{r}$ be other points so that $(x_i,y_i)_{i=1}^{r}$ form a set satisfying condition (2). Then there exists an interpolatory integration rule

$$Q(f) = \sum_{i=1}^{r} A_i f(x_i,y_i)$$

which is precise for all $h \in H$.

We first show that $A_i = 0$, $i = p+1,\dots,r$. Assume to the contrary that $\sum_{k=p+1}^{n} |A_k| > 0$. We construct a $h_\lambda \in H$ satisfying

$$h_\lambda(x_i,y_i) = -(1/\lambda) \quad , \quad i = 1,\dots,p ,$$

$$h_\lambda(x_i,y_i) = \lambda \operatorname{sgn} A_i \quad , \quad i = p+1,\dots,r .$$

Let be λ_o sufficiently large that

$$\iint_B h_{\lambda_o}(x,y)\,dxdy = \sum_{i=1}^{p} (-(1/\lambda_o)A_i) + \sum_{i=p+1}^{r} \lambda_o |A_i| > 0 \quad .$$

It follows from $Z_1 = \{(x,y) \in T \mid h_{\lambda_o}(x,y) < 0\}$ that

$$K: = \inf\{(f-h_o)(x,y) \mid (x,y) \in T \backslash Z_1\} > 0 \quad .$$

Let be $c: = \|h_{\lambda_o}\|/K$. Now we obtain $|(1/c)h_{\lambda_o}(x,y)| \le K$ for

$(x,y) \in T$. It follows

$$(h_o+(1/c)h_{\lambda_o})(x,y) \le f(x,y) \qquad (x,y) \in T$$

and therefore

$$\iint_B (f(x,y)-(h_o+(1/c)h_{\lambda_o})(x,y))dxdy < \iint_B (f(x,y)-h_o(x,y))dxdy \; .$$

Now we obtain a contradiction

$$\iint_B h_o(x,y)dxdy < \iint_B (h_o+(1/c)h_{\lambda_o})(x,y)dxdy \quad .$$

We now show $A_i \ge 0$, $i=1,\ldots,p$. Suppose that there exists an $A_k < 0$. For $\lambda > 0$ let be

$$h_\lambda(x_i,y_i) = -1 \; , \quad i \ne k$$

$$h_\lambda(x_k,y_k) = -\lambda \quad .$$

Then we conclude

$$\iint_B h_\lambda(x,y)dxdy = \sum_{i=1}^{r} A_i h_\lambda(x_i,y_i) = -\sum_{i\ne k} A_i + \lambda|A_k| \quad .$$

For a sufficiently large λ_o we have $\iint_B h_{\lambda_o}(x,y)dxdy > 0$. Arguing as before we find a c such that

$$h_o(x,y) + (1/c)h_{\lambda_o}(x,y) \le f(x,y) \qquad (x,y) \in T$$

and we obtain a contradiction

$$\iint_B (f(x,y) - (h_o+(1/c)h_{\lambda_o}(x,y))dxdy < \iint_B (f(x,y) - h_o(x,y))dxdy$$

proving our theorem.

Example 2 Let H be a space spanned by monomials

$$H = <\{x^i y^j\}_{\substack{i=0,\dots,m \\ j=0,\dots,n}}> \quad ,$$

$f(x,y) = x^{m+1} y^{n+1}$, $T = \bigcup_{i=0}^{n} T_i$ and $m+1$ even.

For the function f we define

$$H_T(f) = \{h \in H \mid h(x,y) \le x^{m+1} y^{n+1} \quad , \quad (x,y) \in T\} \quad .$$

It can be proved that a set of points

$$W = \{(x_{ij}, y_j)_{\substack{i=0,\dots,m \\ j=0,\dots,n}}\} \ , \quad (x_{ij}, y_j) \in T_j \ , \quad i=0,\dots,m \qquad (4)$$

satisfies condition (2). We obtain

THEOREM 4 Let be Q an integration rule of interpolatory type with nodes W of the form (4) and positive coefficients. Then there exists an integration rule $\bar{Q}$ which has $(m+1)(n+1)/2$ nodes with positive coefficients.

Proof We see that $f-h_o$ has at most $(m+1)/2$ distinct zeros on T_i . Therefore $Z(f-h_o) \cap T$ is a subset of a set W satisfying (4). Now we conclude from Theorem 3 the existence of the integration rule $\bar{Q}$ since on the assumptions of our theorem we are able to prove that problem (3) has a solution. $\bar{Q}$ is a minimal point integration rule.

We are also able to extend these results to other function spaces for example spline functions.

REFERENCES

1. Bojanic, R. and DeVore (1966). On Polynomials of Best One-Sided Approximation, *L'Enseignement Math.* 12, 139-164.

2. DeVore, R. (1968). One-Sided Approximation of Functions, *J. Approx. Theory* 1, 11-25.

3. Micchelli, C.A. and Pinkus, A. (1977). Moment Theory for Weak Chebyshev Systems with Applications to Monosplines, Quadrature Formulae and Best One-Sided L^1-Approximation by Spline Functions with Fixed Knots, *S.I.A.M. Jour. Num. An.* 8, 206-230.

4. Strauss, H. (1975). Approximation mit Splinefunktionen und Quadraturformeln, *in* "Spline Functions Karlsruhe" (Ed. K. Böhmer, G. Meinardus, W. Schempp) Springer-Verlag.

PROBLEMS PROPOSED DURING THE SYMPOSIUM

1. (M.F. Barnsley) Given a real-valued function $F(w,z)$ defined on $0 \leq w < \infty$, $0 \leq z < \infty$, and having the properties:

(i) for any fixed $z \geq 0$, $F(w,z)$ can be expressed

$$F(w,z) = \int_0^\infty e^{-wt} d\mu(t) \quad \text{for all} \quad w \geq 0 \quad , \tag{1.1}$$

where $\mu(t)$ is monotone nondecreasing on $0 \leq t < \infty$, and

(ii) for any fixed $w \geq 0$, $F(w,z)$ can be expressed

$$F(w,z) = \int_0^\infty e^{-zs} d\nu(s) \quad \text{for all} \quad z \geq 0 \quad , \tag{1.2}$$

where $\nu(s)$ is monotone nondecreasing on $0 \leq s < \infty$; given also an 'initial' set of coefficients, say

$$\{F_{00}, F_{10}, F_{01}, F_{20}, \ldots, F_{PQ}\} \quad , \tag{1.3}$$

occurring in the formal power series expansion

$$F(w,z) \sim \sum_{m=0}^{\infty} \sum_{n=0}^{\infty} F_{mn} w^m z^n \tag{1.4}$$

of $F(w,z)$ about $w = z = 0$; *determine* the best possible upper and lower bounds on $F(w_0, z_0)$ for given $w_0 > 0$, $z_0 > 0$.

2. (H. Shapiro) Let X be a strictly normed Banach space and Y an n-dimensional subspace. Let S be a translate of an n+1-dimensional subspace of X. Denote by P_Y the metric projection on Y, and let $P_Y^{-1}(0) = \{x \in X: P_Y x = 0\}$.
Conjecture $P_Y^{-1}(0) \wedge S$ cannot consist of exactly one point. (In fact it seems likely that this set, if non-empty, contains a non-degenerate arc through each of its points.)
Fact This is known in case S is a subspace of dimension $\geq n$. (The question originates with B.O. Björnestål.)

3. (H. Shapiro) Let $\mathbb{D}$ denote the open unit disc, and A the set of all complex $f \in L^\infty(\mathbb{D})$ which annihilate the analytic functions, i.e. satisfy $\iint_{\mathbb{D}} f(re^{i\theta}).(re^{i\theta})^n r dr d\theta = 0$, $n = 0,1,2,\ldots$

Let $\phi(z)$ be the function equal to $+1$, $\operatorname{Im} z > 0$, and -1, $\operatorname{Im} z < 0$ $(z \in \mathbb{D})$. Edgar Reich has shown: $\exists\ \delta > 0$ and $f \in A$ such that

$$\operatorname*{ess\,sup}_{z \in \mathbb{D}} |f(z) - \phi(z)| \leq 1 - \delta \quad .$$

(This result arises in certain problems of quasiconformal mapping.)
Problem What about the analogous question for some other ϕ, say $\phi = \pm 1$ on regions separated by a *curve*?

4. (H. Shapiro) Is the following assertion true or false?

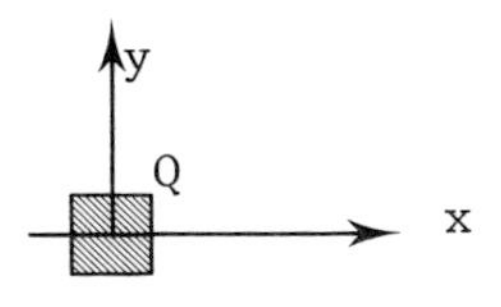

'Q = unit square in $\mathbb{R}^2 \Rightarrow \exists$ constant $C(Q)$ such that, for all smooth f,

$$\inf_{c\in R} \iint_Q |f(x,y)-c|\,dxdy \leq C(Q) \iint_Q \left[\left|\frac{\partial f}{\partial x}\right| + \left|\frac{\partial f}{\partial y}\right|\right] dxdy\ .'$$

Remark This is a Poincaré-type inequality for L^1. The analogous assertions for L^p $(p > 1)$ and max norms are true and well known. I would guess it is true in this L^1 case too.

5. (H. Shapiro) (This question may be trivial.)

Let R be a distribution on $\mathbb{R}$ with the following properties:

(i) $\operatorname{supp} R \subset [0,1]$,

(ii) $R = \mu_1 + D\mu_2$ where μ_1, μ_2 are bounded measures on $\mathbb{R}$.

Is it true that $R = \nu_1 + D\nu_2$ where ν_1 and ν_2 are bounded measures supported in $[0,1]$?

6. (H. Shapiro) Let Ω be a plane bounded domain.

Fact The p.d.e. $\Delta u = e^u$ has a unique solution in Ω which tends to $+\infty$ at $\partial\Omega$. Call this solution u_0.

Conjecture If Ω is convex, the level curves of u_0 are convex.

Remark This has a physical interpretation, that, given a (classical ideal-fluid) flow of a fluid in Ω with a free vortex, the vortex will describe a convex trajectory.

7. (S. Ellacott) The problem of uniform convergence of the method of conformal mapping that I presented can be reduced to the following.

Let S be the open unit circle and $\bar{S}$ its closure. Consider a sequence $\{\theta_n\}$ of functions analytic on S, continuous on $\bar{S}$, $\theta_n(0) = 0$, $\theta_n'(0)$ real and positive.

The sequence $|\theta_n(z)/z| \to 1$ uniformly on $\bar{S}$. It is easily shown that $\theta_n(z) \to z$ on S, uniformly on compact subsets.

Questions

(i) Under what conditions (on θ_n) do we have $\theta_n \to z$ uniformly on $\bar{S}$?

(ii) Is it possible, under any reasonable conditions, to find computable numbers C, α such that

$$|\theta_n(z)-z| \le C\varepsilon_n^\alpha, \text{ where } \varepsilon_n = \max_{z\in\bar{S}}|\theta_n(z)/z| = \max_{x\in\bar{S}}|\theta_n(z)| \text{ ?}$$

8. (J.R. Rice) Consider a region Ω in R^N and the interval $I = [0,1]$. Let $\gamma: I \to \Omega$ be a 'space-filling curve' map which is 1-1 and onto and continuous.

A. There are a number of simple ways to construct γ for simple regions, e.g. rectangles or simplices. Do some of these extend readily to more general domains Ω (i.e. so that they can be easily computed)?

B. Let x_i, $i = 1$ to m, be points in Ω and $t_i = \gamma^{-1}x_i$. Partition I at the midpoints of the intervals $[\tau_j, \tau_{j+1}]$ where τ_j, $j = 1$ to m, is a linear ordering of the t_i. Let w_j be the length of the interval containing τ_j. Then least squares approximation (or other similar norms) on the x_j with the corresponding weights w_j is a good thing to do because it compensates for the scattering (non-uniform spacing) of the x_i. What are the convergence properties of this approximation procedure as $m \to \infty$?

C. The same scheme generates a quadrature rule for Ω; what are its convergence properties?

D. Are there some space-filling curves which have 'better' convergence properties than others? Does it matter whether γ is continuous? (There are some discontinuities which preserve the volume-length relationship and are particularly simple to compute.

9. (D.D. Stancu) Find a n-fold linear interpolation procedure for constructing the Bernstein-type polynomial associated with a function f defined on [0,1] :

$$(P_m^{<\alpha>} f)(x) =$$

$$\sum_{k=0}^{m} \binom{m}{k} \frac{x(x+\alpha)\ldots(x+(k-1)\alpha)(1-x)(1-x-\alpha)\ldots(1-x-(m-k-1)\alpha)}{(1+\alpha)(1+2\alpha)\ldots(1+(m-1)\alpha)} f\left(\frac{k}{m}\right),$$

where α is a non-negative parameter.

Extend this procedure to the bivariate case for obtaining the Bernstein-type polynomials defined at (2.1) and (3.8) in my paper. (See Section 5 of the paper and references [19] and [20].)

10. (D.D. Stancu) It is known that the Bernstein polynomial is a special case of the Schoenberg spline approximation formula (see reference [11] in my paper). How should we extend Schoenberg's approximation method so that the Bernstein-type polynomials considered in the preceding problem remain special cases?

Editor's note: The problems reproduced above were written into a book which was placed for this purpose in the bar. Any response they may have drawn was not committed to writing.